Organizational Innovations

Organizational Innovations

Peter Clark

SAGE Publications
London • Thousand Oaks • New Delhi

First published 2003

Apart from any fair dealing for the purposes of research or private study, or criticism or review, as permitted under the Copyright, Designs and Patents Act, 1988, this publication may be reproduced, stored or transmitted in any form, or by any means, only with the prior permission in writing of the publishers, or in the case of reprographic reproduction, in accordance with the terms of licences issued by the Copyright Licensing Agency. Inquiries concerning reproduction outside those terms should be sent to the publishers.

SAGE Publications Ltd
6 Bonhill Street
London EC2A 4PU

SAGE Publications Inc.
2455 Teller Road
Thousand Oaks, California 91320

SAGE Publications India Pvt. Ltd
32, M-Block Market
Greater Kailash – I
New Delhi 110 048

British Library Cataloguing in Publication data
A catalogue record for this book is
available from the British Library

ISBN 0 7619 5881 9
ISBN 0 7619 5882 7 (pbk)

Library of Congress control number available

Typeset by Mayhew Typesetting, Rhayader, Powys
Printed and bound in Great Britain by Athenaeum Press,
Gateshead

Contents

Preface

This book arises from a stream of research, largely funded by the Economic and Social Research Council (ESRC) and the Engineering and Physical Sciences Research Council (EPSRC), carried out over the past two decades. Our five-year programme grant for the Work Organization Research Centre at Aston University focused upon the 'Longitudinal and comparative investigation of technology and organization' (Peter Clark, John Child, Ray Loveridge and Jennifer Tann with Richard Whipp, Chris Smith, Margaret Grieco, Ken Starkey, Alan McKinlay, Neil Staunton, Mick Rowlinson). My contributions were published in three books: *Innovation and the Automobile Industry* (with Richard Whipp 1986); *Organization Transitions and Innovation Design* (with Ken Starkey 1988) and my *Anglo-American Innovation* (1987). This programme was both a pillar in the Midlands Organization Behaviour Group, including Warwick (Andrew Pettigrew), Keele (John Hassard), Leicester and Nottingham, and also a focus for many stimulating visitors: Howard Aldrich, Bill McKelvey, Bill Scott and returns to Aston by Stewart Clegg and Lex Donaldson. John Child has been a friend and colleague for a long time. We do have quite different ontological inclinations and for me those differences have been valuable as stimulus and as clarification.

The Work Organization programme led into a clutch of projects undertaken jointly with colleagues in the Operations Management area, particularly David Bennett and Peter Burcher. The focus became the new technologies for co-ordinating firms and networks through information technology, and the challenges created by organizational innovations attributed to Japan. The initial architecture of those studies was set out in *Innovation in Technology and Organisation* (with Neil Staunton 1989). This programme soon came to involve Sue Newell (now Royal Holloway) and Jacky Swan (Warwick) whose contributions extended the original programme. Sue and Jacky then moved to Warwick where they were central in establishing 'Innovation, Knowledge and Organization Networks' (IKON) with Mike Bresnen, Maxine Robertson and Harry Scarbrough (Leicester). IKON provided a key hub for an array of programmes. My research benefited from collaboration with Angela Dumas (London Business School) and Peter Hills (EPSRC Design Consultant) in an EPSRC study of the role of totem-building and metaphor in organizational innovation and technology pools.

The references to the European Innovation Monitoring Systems (EIMS) refer to a stimulating European Union (DG XIII) project involving six nations skilfully led by Bengt Stymne (Stockholm). The British group involved Frank Mueller (Leicester) and Fred Steward (Aston). The other European nations were France, Italy, Spain and Germany. We also had American, Japanese and Canadian colleagues, including Harvey Kolodney (Toronto). This project allowed me to investigate the role of quasi-market mechanisms in the American and European approaches to health provision. This mixture of colleagues and nations provided a useful challenge to existing research and theorizing.

The attention to the geography of innovation and to some aspects of consumer research arose from a working group at Birmingham University, involving Nick Henry (Newcastle), Jane Pollard, Isabelle Szmigin and Kirstie Ball, which was formed to explore the theme of knowledge, consumption and organization geography. Nick and Jane were constructive proponents of the geographical imagination.

My special thanks are due to those in the IKON network and its peripheries. Jacky Swan, Sue Newell, Harry Scarbrough and Maxine Robertson have generously shared their sense of excitement, choice of directions and working papers. Although not in the core of IKON, my colleagues Chris Carter (St. Andrews) and Frank Mueller (Leicester) provided a rich sense of alternative explanations and the relevance of irony. Candy Jones (Boston College) accepted an invitation to visit our ESRC seminar series where she gave an insightful and thoughtful account of the longitudinal perspective. The longitudinal issue has also benefited greatly from discussions with Mick Rowlinson (North London).

My aim in this book is to carry the themes of my *Organizations in Action: Competition between Contexts* (Clark 2000) more fully into the field of innovation studies. My view is that there are a number of informative accounts of earlier theorizing of innovation, especially the economics of innovation, but that these underplay recent developments.

My special thanks to the University of Birmingham for the support I received in researching and writing this book. Particular thanks to Jane Whitmarsh for the artwork. Finally, and most importantly, my wife, Jennifer, has cheerfully lived through another book project.

Peter Clark

Acknowledgements

We are grateful to the following publishers for permission to reprint and develop previously published material:

Blackwell Publishing Ltd
Gregory, D. (1994) *Geographical Imaginations*, pages 401 and 412, Figures 30 and 31.
Maguire, J. (1999) *Global Sport: Identities, Societies, Civilizations*, page 80, Figure 3.
Swan, J., Newell, S. and Robertson, M. (1999) 'National differences in the diffusion and design of technological innovation: the role of . . .', *British Journal of Management*, 10(3): S53, Table 3.

Cambridge University Press
Archer, M.S. (1995) *Realist Social Theory: The Morphogentic Approach*, page 193, Figure 10.

Elsevier Science
Child, J. (2000) 'Theorizing about organization cross-nationally', in *Advances in Comparative Management*, 13, page 58, Figure 3.
Clark, P.A. (1985) 'A review of the theories of time and structures for organizational sociology', *Research in the Sociology of Organizations*, 4, page 44, Figure 1.

MIT Press
Bijker, W.E. (1995) *Of Bicycles, Bakelites, and Bulbs: Toward a Theory of Sociotechnical Change*, page 53, Figure 2.13.

National Academy Press, Washington, DC
Kline, S. and Rosenberg, N. (1986) 'An overview of innovation', in R. Landau and N. Rosenberg (eds), *The Positive Sum Strategy*, by the National Academy of Sciences, page 290, Figure 3.

Routledge
Clark, P.A. (2000) *Organizations in Action: Competition between Contexts*.
Ackroyd, S. and Fleetwood, S. (eds) (2000) *Realist Perspectives on Management and Organisations*, page 100, Figure 5.1.

Sage
Clegg, S.R. (1989) *Frameworks of Power*, page 214.

M.E. Sharpe, Inc., Armonk, NY
Clark, P., Newell, S., Swan, J., Bennett, D., Burcher, P. and Sharifi, S. (1992–93) 'The Decision Episode Framework and Computer-Aided Production Management (CAPM)', *International Studies in Management and Organization*, 22, 4, page 72, Figure 1.

The author would also like to thank Candace Jones for access to her working paper 'Strategic networks in American film, 1895–1920' (2000).

The New Political Economy Agenda 1

T he aim of this book is to provide an actionable understanding of organizational innovation for students and for practitioners. The extensiveness of organizational innovation is illustrated by examining the period 1950–77. In that period Pascale (1990: 20) finds 27 examples of organizational process innovations that emerged, became known and were then viewed as just one of many options. The listing included the following: Decision Trees, Managerial Grid, Job Enrichment, Theory X & Y, Brainstorming, T Groups, Management by Objectives, Diversification, Experience Curve, Strategic Business Units, Zero-based Budgeting, Value Chain, Decentralization, Quality Circles, Restructuring, Portfolio Management, Management by Walking About, Matrix, *Kanban*, Intrapreneurship, Corporate Culture, One-Minute Manager. Since 1977 many more organizational innovations have been commodified and sold by specialist suppliers. The sales are to niche markets of eager managers in firms of every kind in almost every nation. Consequently, there are now a whole series of specialist reviews by consultancies of the rise and fall of new process innovations (e.g. Business Process Re-engineering). There is a growing academic debate over whether these are simply fashions or whether they really add to the effectiveness of organizational performance (e.g. Abrahamson 1991; Kieser 1997). For example, Scarbrough and Swan (2001) contend that Knowledge Management (KM) emerged around 1995 and peaked at the new millennium.

Underlying the above listing of seemingly transitory fashions there are three core processes of transition in capitalism which are driving the concern with organizational innovation and the search for a more useful innovation theory:

(1) Process innovations are used by consultancies and professional associations to market their services and to commodify their products.
(2) The demand for professional services has been driven by the shift in all kinds of firms away from mindsets based on structural solutions (e.g. organization charts, formalizing procedures, creating specialist roles, creating new divisions) to processual solutions and mass customization (see Chapter 2).
(3) There have been massive developments in the process technologies of information management, surveillance, remote management, closed-circuit TV, computer-aided engineering and design and similar.

These have altered the approach to innovation by organization theorists, geographers, marketing and information systems specialists.

This book aims to update the notion of a useful theory of innovation outlined by Nelson and Winter (1977) through the analysis of organizational innovations as hybrid networks of knowledge, power and technology. The opening chapter is a stylized piece of scene-setting relevant to the new agenda on innovation and knowledge. The style of the chapter is one of broad contrasts and some necessary polemic. The next section links the dispersed political economy of market capitalism with the notion of performative knowledge. Then the contrasts between the old agenda and the emerging agenda are highlighted. The contrasts are unpacked by examining a major shift in ways in which innovation is analysed. The final section explains the selected themes and the structure of the book.

Old agenda, new agenda

The section contrasts the old agenda with the new agenda. The contrast is between a coherent and respected orthodoxy stretched to the limits of its analytic relevance in the old agenda and an exciting, demanding new agenda that is more loosely formed and is still emerging and developing. The old agenda was formulated in the 1945–70 era of Big Science, modernity and late modernity. The old agenda focused upon the *structure of technology innovation* and the new agenda focuses upon the *processes of organizational innovation*. The contrast reflects a deep change in the formative background assumptions that guide the theory and practice of innovation.

Figure 1.1 amplifies the main theoretical schools from the social science and economics that are involved in understanding and managing innovation (Clark 2000). The modern, the anti-modern and the new political economy of process are shown stylistically in Figure 1.1. From the perspective of the new millennium, late modernity is a curious mixture of beliefs in reason and utopian romanticism. Those taken-for-granted assumptions (Searle 1995) were vigorously challenged and caricatured by the various anti-modern schools. They successfully demonstrated the limits of the modern and established significant new views about innovation. The anti-modern sought to demonstrate the necessity for examining innovation as a multi-level, dynamically recursive and processual configuration (see Chapter 2) rather than a linear sorting of variables (Mohr 1982). These processual perspectives formed the basis of syntheses (e.g. Van de Ven and Poole 1995) and many others (e.g. Clark and Staunton 1989). In the past decade the debate between the modern and the anti-modern has been joined by an emergent position: the *new political economy of process*. The three layers of Figure 1.1 span the transition from the 'late modern' through the anti-modern and then into the new political economy (Alexander 1995). These three layers chart a search for a process-oriented

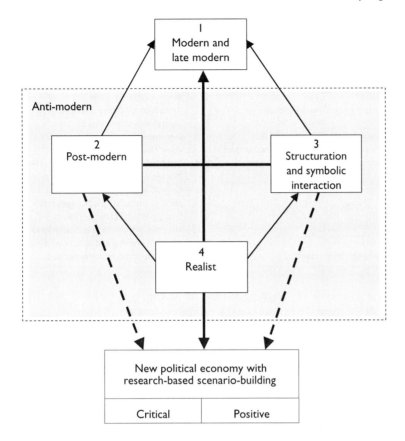

FIGURE 1.1   Process-oriented framework (developed from: Clark 2000: Figure 15.1)

framework in useful innovation theory. This section concentrates upon the contrast between the modern and the new political economy.

Four main areas are examined to contrast the old and the new agendas:

- Useful process theory
- Global political economy and nations
- Firms-in-context
- Innovation characteristics.

These are summarized in Tables 1.1, 1.2, 1.3 and 1.4. The intention is to problematize the persisting background assumptions and tired dichotomies that have mired the development of useful innovation theory (Nelson and Winter 1977; Mohr 1982).

First, in the old agenda (see Table 1.1) theory aimed to construct deterministic, law-like statements about the efficient firm [1]. Equilibrium models promoted efficiency and therefore had to treat innovation as a special event that shifted the equilibrium [2]. Efficiency was the dominant assumption. Research and development investment [3] and in technology

TABLE 1.1   Critical problems in process theory

| Process theory | |
Old agenda	New agenda
1. Deterministic law-like theory.	1. Meta-frameworks for understanding. The structure/agency issue.
2. Efficiency regulates innovation.	2. Innovation/efficiency dilemma. Knowledge inputs/innovation outputs.
3. Research & Development (R&D) expenditures. Technostructures at core.	3. Innovation-and-design. Whole firm and its value chain.
4. Technology and production orientation.	4. Production and consumption interwoven. Market and consumer society.
5. Time–space.	5. Uneven change in multiple times. Longer periods (decades). Path dependency. Sticky periods. Exits and failures. Geo-historical specificity and institutional theory. Long-run and short-run dynamics.
6. Universal design rules.	6. Research-based scenario-building.

and raw materials [4] were presumed to be the driver of successful innovation. Therefore there was considerable research on the readily available data about investment (inputs) and profitability (outputs) with slight attention to the intervening process of throughput. It was here that Galbraith (1967) explained the role of the technostructure [3], consisting of graduate, salaried scientists and professionals who linked firms to the university research centres. They provided the reflective and reflexive monitoring (Giddens 1984) on the present and sought to anticipate the future (Perrow 1967). These models were oriented to production and technology rather than consumption [4]. Even the pioneering market orientation of Project Sappho led by Chris Freeman became marginalized. Academic research on innovation was strongly shaped by the modernist conception of an objective knowledge produced by experts in special centres. This law-like knowledge was constructed in a format that was abstracted from time–space [5]. Specialist experts transposed the knowledge into universal design rules [6] for use by government and in firms.

The old agenda concentrated upon the structural features rather than the underlying processes. Exponents of this agenda frequently observed that it was only possible to have knowledge about the structure and not about the complex, fluid processes (e.g. Pugh and Hickson 1976).

In the new agenda for process theory the critical problem is how to characterize recursiveness and processes in multi-level, contingently dynamic configurations and to distinguish between their reproduction over successive iterations and their transition to a new form [1]. The problem is still not yet neatly formulated. This lack of formulation is expressed in the debate over structure and agency. The debate explains the role of the structuration theory of Giddens in organization theory (e.g. Barley, Orlikowski), in information systems (DeSanctis and Poole 1994), in geography (Gregory 1994) and in innovation studies. This core issue is examined fully in Chapters 2 and 3. In the new agenda, deterministic

TABLE 1.2 Critical problems of political economy and nations

Political economy	
Old agenda	**New agenda**
1. Slight political economy.	1. Global competition between contexts. Capitalism colonizing everyday life.
2. Single developmental capitalism.	2. Multiple capitalisms.
3. Nation in background.	3. Nation in foreground. National systems of innovation. Societal contingencies. Social capital theories.
4. Mass production and distribution.	4. Hypothesis of mass customization as the leading edge.

theory is replaced by an eclectic theory containing many frameworks, each with specialist tasks. There are meta-frameworks that cover specific issues. The relationship between innovation and efficiency [2] is expressed as a continuing dilemma (Abernathy 1978; Clark and Staunton 1989). Also, knowledge becomes important as input to the whole firm [3]. This is expressed in the notion of innovation-design (Clark and Starkey 1988; Clark and Staunton 1989) rather than Research and Development. Consequently, analytic attention embraces consumption and the consumer society [4] rather than simply production and technology. Now there are critical problems of understanding, expressing and framing the temporal and spatial permanencies and differences [5]. Attention is given to the short- and long-run dynamics. Attempts are made to chart uneven, surprising change (e.g. Abernathy and Clark 1985). There is attention to sticky periods in which established routines are difficult to alter. Also, the population ecology of firms is subject to creative destruction. The notion of social trajectories and of long-run path dependency is both widely used and is increasingly a focus on critical revision. In the new agenda the critical issues are still in development and are the subject of debate.

Second, the critical problems of political economy and the nation are shown in Table 1.2. In the old agenda there was slight attention to the international political economy [1]. Consequently there is slight attention to how international markets reflect the unequal power in the world economy (see Chapter 5). Market capitalism and the so-called 'end of history' (Fukuyama 1992) is a western-dominated arrangement within which the USA still occupies an exceptional position. In the late modern era there was an assumption that in capitalism the firm and the nation evolved down a single developmental path [2]. Issues of place (e.g. Bombay and San Francisco) were relegated into abstract notions of space and so the nation as an analytic entity was always in the background [3]. The late modern era was the pinnacle of mass production and of mass distribution [4] (Chandler 1977; Harvey 1989; Dicken 1998).

In the new agenda there are many critical problems. There is an assumption of global competition between contexts [1] so that the place where you are and its geography matters (e.g. Landes 1998). So far Bombay is less of a centre for electronic embrace than is San Francisco.

TABLE 1.3   Critical issues of the firm-in-context

Firm-in-context	
Old agenda	**New agenda**
1. Independent, agentic.	1. Co-evolution of firm and context. Elective affinities thesis.
2. Research and Development. Left-to-right models.	2. Innovation-design. Parallel interactive models.
3. Technology artifacts.	3. Organizational network innovations.
4. Inter-firm nexus dormant.	4. Firms in clusters, webs and chains. Pivotal firms and their webs.
5. Structure and contingency.	5. Finite capabilities, learning and knowledge.
6. Learning-by-doing. Change through adaptation.	6. Adaption/selection dilemma. Change by population ecology.

Location matters (Dunning 1993). Porter (1990, 1998) contends that the home base of the firm and its chosen markets constitute a structure that significantly shapes the zones of manoeuvre available to local agents (see Chapter 5). The uneven character of national market power is made transparent through post-colonial perspectives (e.g. Said). Transparency decodes the role of professional models of management in articulating the link between the state and the professions in key advanced economies (e.g. USA, Germany). These professional models are presented as politically neutral in Organization Theory. However, in the new political economy their discursive cloak of rationality, progress and reality is used by the west to disseminate knowledge embedded in uneven power. There are multiple capitalisms (e.g. North American, European, Asian hybrids) with many sub-varieties [2]. The nation is in the foreground [3], but the international political economy and the differences in power between nations inform the analytic position of the nation. Consequently, enormous attention is given to national innovation-design systems (NIDS) and to societal contingencies (Clark and Mueller 1996). There is a strong interest in societal analysis and the significance of social capital within nations. In the new agenda the political economy is hypothesized to contain a significant and growing proportion of mass customization with short design cycles and short product life cycles [4] (Pine 1993).

Third, the firm-in-context shown in Table 1.3 is radically different in the new agenda. In the old agenda the firm is largely treated as being singular and as possessing the agency necessary to alter its position through strategic choice by its management [1]. Management invests in Research and Development [2] by drawing upon research undertaken in the universities. There is a linear flow from basic research into its application in the firm. The outputs of this process are technological artifacts [3]. Although firms were actually in networks, those inter-organizational networks were largely ignored and their crucial role in the division of knowledge was obscured [4] (Clark and Staunton 1989: Ch. 8). The organization was analysed as a structure [5] rather than a process. Structure could be readily described by reference to observable features such as charts, titles, names of departments and an array of proxy variables. Much of this empirical theorizing epitomized positivism. Moreover, the findings were interesting but of limited use. Little was learnt about the processes of innovation. The notion of learning-by-doing [6] provided some insights, but required considerable development. There was an implicit acceptance of the romantic enlightenment notion that firms should and could achieve change through adaptation.

In the new agenda the critical problems are many. The aim is to theorize firms as being in a process of co-evolution with their contexts [1]. Firms and contexts should aim to possess those enabling characteristics that provide an elective affinity to successful trajectories of profitability. The notion of innovation-design [2] as a total, long-term process is used to frame both efficiency and innovation. Firms are presumed both to possess valuable knowledge and also to know as much about the critical problems as is known in the universities. Left-to-right models are refuted and replaced with interactive models with many parallel activities. Moreover, the nature of linkages between firms and universities varies over time. Organizational network innovations [3] are of major interest. Firms are analysed within clusters, webs and chains [4]. These networks frequently possess hubs and pivotal firms that orchestrate activities and allocate roles (e.g. Hakansson 1988; Ford et al. 2000). Attention to process [5] is expressed in a variety of notions, including capabilities and the resource-based theory of the firm. Heavy attention is given to knowledge and learning. One of the key issues is just how far the capabilities are finite and to what extent learning is specific, idiosyncratic and heterogeneous (Nelson 1991). The old agenda presumes that learning-by-doing promotes the adaptation of the firm to new circumstances [6]. The new agenda indicates that existing firms face the 'winds of creative destruction' and may therefore exit from the market. Only some firms may survive.

Fourth, conceptualizing innovation is summarized in Table 1.4. In the old agenda there is a sharp dichotomy between 'normalcy' and epochal innovation [1]. The context is given minimal attention [2] while attention is focused upon the drama of innovation. The characteristics of the innovation are treated as though they are framed at a particular moment and remain stable for very long periods. The features of the innovation and its context (e.g. the design hierarchy) remain unknown. Best practice [4] is singular. Innovations, once discovered, are diffused through centre–periphery models [5] by a hidden though very powerful set of supply firms [6]. Power and political process [7] was previously obscured, but now the process of innovation is theorized as a contested arena. Within the arena both the powerful and the less powerful may try to smuggle in their preferences and to outflank one another. In the case of drug firms, their influential role is well known. Yates (1989) shows that American firms supplying systems and artifacts for control through communication are also very influential, yet their roles remain silent in the supply-oriented frameworks. With few exceptions the politics of innovation and relative power of different players are brilliantly ignored. Leading texts illustrate –

TABLE 1.4   Critical problems in conceptualizing innovations

Conceptualizing innovations	
Old agenda	**New agenda**
1. Dichotomy of epochal/'normal' innovation.	1. Continuous with peaks. Emergent phenomena. Discontinuities.
2. Minimize context.	2. Context heavily implicated.
3. Stable characteristics.	3. Emergent hybrids.
4. Best practice.	4. Multiple scenarios of best practice.
5. Centre–periphery models.	5. Interactive and pluri-centre models.
6. Supplier domination.	6. Users play key role. Multiple groups.
7. Power obscured. A-political.	7. Politically contested, smuggled.

without a hint of irony – how in Third World nations the introduction of western innovations often transform the social structure (Appadurai 1986a).

In the new agenda innovation is so extensive and so modulated that an array of analytic frameworks have been and are being developed to conceptualize the innovation process [1]. Innovation is acknowledged to be continuous with peaks. There is the expectation of surprise and the recognition of emergent, unintended outcomes. The context of innovation is now integral to analysis [2]. Also any innovation is presumed to be contingently stable with varying degrees of interpretative flexibility. Hybrids [3] are commonplace, even normal (Abo 1994). Consequently there are multiple forms of best practice [4]. Anticipating the future is no longer undertaken through design rules. There is increasing use of scenario-building. Interactive models that recognize the potential existence of multiple centres are replacing the dominant centre–periphery models [5]. Also, the powerful suppliers [6] are challenged and even usurped by the users of innovations. Users play a majestic role in the innovation process. The politics of innovation [7] now acknowledges contest, smuggling, rejection, boycotts and the appropriation of innovations (see Szmigin 2003). The definition of the firms (see Chapter 9) has altered to accommodate these many political currents.

In the new agenda innovations are conceptualized as dynamic bundles of elements whose bundling may be reconfigured at any moment by either/both suppliers and users. The role of management information systems is the simple example. The specificity of innovation is dynamically contingent (Clark and Staunton 1989: Ch. 3). Hybrids abound. American Football is a hybrid from American folk games, rugby union and soccer imported from England (Clark 1987). Of course, there is widespread failure. Perhaps two-thirds of attempts fail and not just in Business Process Re-engineering (Hammer and Champy 1993). We should analyse innovations as an ensemble whose fusion depends on complementarities and also possesses limits (Gille 1978). The earlier romanticism about successful innovation is transformed by concepts of zones of manoeuvre and proximal development. Innovations are increasingly systemic and are articulated through knowledge and knowing (Cook and Brown 1999). The problem of constituting large-scale systems of innovations is immense (Hughes 1983, 1990).

Core themes

The book highlights selected focal concepts in the process-oriented framework for innovation of Figure 1.1. Intra-firm and inter-firm innovations are examined within their local, institutional and global context (i.e. a multi-level framework). The international political economy and the consumer society are integrated and explained. The focus is upon a particular form of organizational and technological innovation for the co-ordination and integration of the firm and between firms. These are the core examples of the nexus of money–time–space that are generic in information systems thinking and practice. Their evolution reflects key contextual shapings. Previous work on innovation diffusion focuses excessively on the supply side (e.g. IT suppliers). This book addresses both the suppliers and the users as well as quasi-autonomous professionals who play an important role in the innovation diffusion and adoption. This book also spans the four processes of diffusion, design, implementation and utilization that are central to organizational innovation and to the development of information technology. Previous works have tended to focus too much on the implementation of pre-defined 'best practice' innovations, thereby neglecting both earlier episodes relating to innovation design (e.g. agenda formation and selection) and the actual uses of organizational innovations.

Six core themes provide integration across the individual chapters. First, *organizational innovations* will be analysed as networks in which technology is a process with an electronic embrace. This theme addresses the central role of information technology. Co-ordination and integration technologies will be emphasized. These have evolved massively over the past two decades and we expect them to continue to evolve over the next decade. Particular attention will be given to innovations in which time, space and resource costs are key features. The time–space dimensions involve both compression as well as stretching over larger spaces (e.g. cloning across firms or regions) or through longer periods (i.e. past, present, possible futures). These organizational innovations increasingly involve uneven bursts of activity of different kinds.

Second, there has been and continues to be a *massive evolution of the international supply side* of hardware, software and consulting firms. Contrary to some accounts of organizational innovations, there does exist a strong, significant and growing international supply side for organizational innovations. International specialist firms play a major role in distributing organizational innovations. In some cases this role has a long history (e.g. McKinsey). There are recent and now major players, such as SAP, the German specialist supplier of financial, accounting and Enterprise

Resource Planning systems. This supply side has in many cases replaced the host firm as the diffuser of the innovation even within the same host firm. This diffusion process in some instances involves a strong American shaping which has considerable consequences for specific European nations. The European Commission has sought to address this competition through initiatives such as the European Innovation Monitoring System (EIMS). Most of the innovation literature and practice is concerned with the supply side interests in their various forms in promoting more rapid diffusion of innovation. This has generated a pro-innovation bias in the literature. This bias is recognized and is addressed in this book.

Third, the *users* have been the silent partners until the past decade. The position of the innovation user has been under-developed. Users are frequently faced by an array of organizational innovations, many of which seem to be good solutions. However, there is often a degree of fuzziness about what the innovation is and the uses to which it can be put. This theme is organized around the Decision Episode Framework in Chapters 7, 8 and 9. Users face the problem of appropriating the organizational innovation. Appropriation is a key process in which the members of the using organization seek to author forms of the organizational innovation which select the architecture and modules which best match the users' situation. The fuzziness of organizational innovations may require the development of special languages for decoding the gap into which the innovation has to be introduced. Technology suppliers play a role as 'fashion setters' who develop rhetorics that push certain innovations to the forefront of management practice. However, users vary in their awareness of the significance of their investment in languages of action. We show that the firm-specific expertise will vary in its capacity to undertake this process of appropriation. The encounter between the supplier and the user provides the opportunity for the co-authoring of innovation, although, as we have indicated, the supply side is often more oriented to, and experienced in, handling the episodes that cumulate into a process of utilization. Authoring enables the successful utilization of the organizational innovation.

Fourth, as already indicated, there are *interpenetrating and overlapping players and layers of analysis*. For example, the suppliers and users overlap and interpenetrate. They are not separate because the professional associations have memberships from suppliers and users and consultants have often had prior experience as users. Research shows the roles of key individuals in spanning between supply and usage. For example, key personnel, with successful histories of implementation of innovation in their CVs, move into the supply side for varying periods. As individuals move between supply and usage their interests in the innovation process may change. For example, users may be interested in designing customized software solutions whereas software suppliers may be interested in a fixed menu of possibilities.

Fifth, *knowledge utilization, appropriation and articulation*. Innovation usually involves the creation, articulation and communication of knowledge, not just of information transfer. The ways in which knowledge is articulated by particular individuals or groups and communicated across social boundaries may vary across different episodes of innovation process across different national or institutional contexts.

Sixth, *national specificities* and *elective affinities* between nations and types of innovation are of increasing relevance. Recent attention to the national level has corrected earlier neglect. The theme of national systems of innovation suggests specificities, yet the issue is the degree of shaping, the factors involved in shaping and the interpretative flexibility. The elective affinity thesis is that without an affinity between national predispositions and the strategic form of the innovation there will be great potential for failure. These arguments are illustrated from current, comparative research. Attention is drawn to the autonomous influence of the involving organizational innovation that can migrate with different effects depending on its immanent, unintended features and their interpretative flexibility. The USA and Japan will be used for illustrative purposes, as well as many other examples drawn from Europe.

Structure of the book

Chapter 2 starts by examining the shift to the new political economy. The focus is upon the hypothesis that mass customization and the post-modern condition have been the main tendencies since the 1980s (see Harvey, Clegg, Jameson, Pine and Giddens). Capitalist firms introduced a decade of Schumpeterian creative destruction with the exit of many established forms of business (Freeman, Dosi). Exnovation rivalled innovation (Clark and Staunton 1989). Moreover the tenets of positive political economy were deployed to ablate whole sections of the managerial service class from their 'safe' lives in welfare capitalism. Business Process Re-engineering (BPR) was just one process technology that reformulated orthodox Scientific Management. BPR assembled new concepts of space and time grounded in flows, movement and complexity theory and with 'process' everywhere. That era also experienced the insertion of massive new potentials for the 'gaze' that was inscribed in the information and communication technologies. This chapter robustly interprets Lakatos's notion of research programmes to examine the three research programmes on innovation: the modern, the anti-modern and new political economy. Castells' analysis of networks as hybrids completes the introduction to the mass customization tendency.

Chapter 3 focuses upon space, time and process. Space–time has been transformed in the past two decades through their stretching, commodification and infusion with the new process technologies (Giddens, Harvey, Castells). These are the social made durable. The approach of heterogeneous engineering is examined (Callon, Latour, Law, Hughes, Bijker, Woolgar). The generative role of knowledge and knowing (Brown, Cook, Wenger, Lave, Weick) is explored through a comparison between the assemblages that built cathedrals without maps (Turnbull) and the modern laboratory where knowledge (not knowing) is inscribed in circulating texts. Control over the knowledge and knowing about the space–time framing of processes and of organizational innovation is a central activity for the global service class. Their gaze on the specifics of place and temporal rhythms transforms those rhythms (Lefebvre, Gregory, Foucault). That transformation is central to the hypotheses of mass customization and translates the post-modern condition. Their Foucauldian gaze enabled the transformational organizational innovation of the shopping mall and stimulated capitalism to use the fruits of Cold War technology to spread systems of surveillance everywhere, both through closed-circuit television (CCTV) and the time–space trajectories of the data trails left by the consumers' everyday usage of credit (Lyon). Giddens' contribution is a major consideration and this is robustly scrutinized to enter a series of revisions. These provide the introduction to the next chapter.

Chapter 4 addresses the problem of moving from the approach of variables to that of multi-level relational configurations (Mohr, Van de Ven and Pettigrew). This starts with the issues raised by the anti-modern schools: the realists, structuration theory, post-modernism (Jameson, Baudrillard) (see Figure 1.1). The new political economy draws eclectically from each of these by using the realist perspective (Harré, Archer) as an analytic cutting-edge. Core elements of the post-modern critique are retained. Two interwoven issues are dominant: first, that of structure and agency (Giddens, Barley, Archer); and second, the issue of how to theorize action as recursive while handling the dynamics between stasis (reproduction) and morphogenesis (transformation). A framework is developed and explained.

Chapter 5 is about the role of global context (Child) and the degree of variation in organizational innovation found in national contexts (Clark, Mueller, Sorge). Location matters and there is competition between globally spread contexts. We construct a reformulated eclectic theory from the frameworks of Dunning, Porter and Rugman. This emphasizes the role of both the national institutions and the market place on domestic firms. In the Cold War many Japanese firms came to regard the USA as their market place (Dower, Cusumano). At the national level there is the isomorphism hypothesis (Scott, DiMaggio, Powell, Oliver) that alleges homogeneity between firms arising from certain causal mechanisms. The isomorphism hypothesis is a classic example of simple contagion theory and leaves too much unexamined. It is countered with the concept of isolating mechanisms (Jones). These define the zones of manoeuvre in space–time that firms are able to construct. Isolating mechanisms draw attention to the heterogeneity of firms. The chapter contains an examination of global corporations.

Chapter 6 examines the market society and conventions of co-ordination for market-production systems. The examination of the market compares orthodox economics with Braudel, Latour, Baudrillard, Jameson and the cultural turn of Miller and Appadurai. Markets are conceptualized as theatres of consumption, but housewives are hardly the vanguard of history (see Miller). Markets contain both commodities and tournaments of value (Appadurai). Mass customization of consumption introduces the space–time issue. Storper and Salais (1997) use conventions of co-ordination to link markets and production.

Chapter 7 continues the market theme by showing the link between Rogers' diffusion of innovations and the powerful gaze of the suppliers of process technologies and organizational innovations upon their customers. Rogers has assembled and edited the seminal account of diffusion, yet, as will be apparent from the previous chapters, this does not explain organizational innovation. The suppliers' position has to be nested in a configuration with the 'pool' of innovations and their users as in the Decision Episode Framework (DEF).

Chapter 8 continues to unpack the Decision Episode Framework with an insightful examination of the 'organizational innovation'. Process technologies are shown to be hybrids of hybrids. We examine the life of several innovations. This chapter draws selectively upon the heterogeneous engineering perspective (e.g. Law, Woolgar, Bijker) and the duality perspective on technology and organization (e.g. Barley, Orlikowski, Yates). We retain an interest in those core features of innovation that relate to corporate performance (Best, Abrahamson, Kogut). We show that innovations are contextualized and frequently differ from the rhetoric of the suppliers (Burcher, Clark).

Chapter 9 examines the general role of firms and the particular position of firms using organizational innovations and process technologies. Users make organizational process innovations (Castells). The Decision Episode Framework presents the user as potentially very active in the process of innovation. Firms contain capacities that are finite and are located in finite zones of manoeuvre. Their antecedent absorptive capacity is a major explanation of the shape and uses of innovation. Castells' notion of articulation is interpreted with particular reference to the boundary-spanning role of professional associations (Newell, Scarbrough, Swan).

Chapter 10 returns to the issue of global transfer and national specificities. Few innovations can stretch space–time across many national boundaries. Two examples are in global media sport and the typical North American/European supermarket with its global commodity chain. The pressures of mass customization require that global firms become processes and that their services/products are variable in different national contexts. The shift in McDonald's illustrates this tendency. These are very distinctive space–time systems. Most innovations are translated when crossing national boundaries, especially between North America and Europe, and also in the translation of Japanese innovations (Abo) by their global firms (e.g. Toyota, Fujitsu). This process illustrates the analytic relevance of notions such as elective affinities, zones of manoeuvre, relational configurations, isolating mechanisms and the revised account of strategic choice (Child).

In the last decade the understanding of organizational innovation and of process innovations has been transformed. There is a new agenda. Its emerging research programme is the best place and temporality from which to start.

Knowledge, Power and Information Technology 2

This longish second chapter contains four themes. First, an account of the hypothesis that we are in a world of mass customization that emerged two decades ago and now contains multiple, different forms of innovation. Second, it is argued that in order to understand the contemporary world of work it is necessary to understand and compare three quite different discourses about innovation. Each is explained and two are characterized as declining, possibly degenerating. A third, the new political economy, is emerging and requires both new assumptions and a new A–Z of key concepts. Third, knowledge, power and information technology are the hybrid networks of Castells' informational capitalism. Fourth, knowledge, power and information technology are implicated in the strategic co-ordination of firms and between firms.

The mass customization hypothesis

Decoding this phase of capitalism occupies the minds of theorists and journalists of every persuasion. The notion that capitalism continuously changes its face while retaining underlying tendencies is hardly contentious. So, how might we characterize the past decade and this decade?

Figure 2.1 draws upon the geo-historical perspectives of Harvey (1989) and Gregory (1994). The figure traces changes in the modes of representing power-knowledge and the regime of financial accumulation. There are three time periods and the central period that separates 1945–75 from 1985–2010 is characterized in terms of the fiscal crisis of capitalism (1975–85). In the period 1945–75 the mode of knowledge-power representation was in terms of disciplines and design rules. Its mode of accumulation was welfare capitalism as the form of regulating mass production and distribution. Capitalism was concentrated in metropolitan areas and their zones of production.

The fiscal crisis of 1975–85 created a crisis for analytic languages and for accumulation. Harvey contends that there was an epochal shift in the compression and commodification of time and space. The crisis was resolved through transforming the previous moral economy of time and space. For example, in retailing, the key innovations were the American shopping malls in the suburbs and outskirts of cities. These reduced their

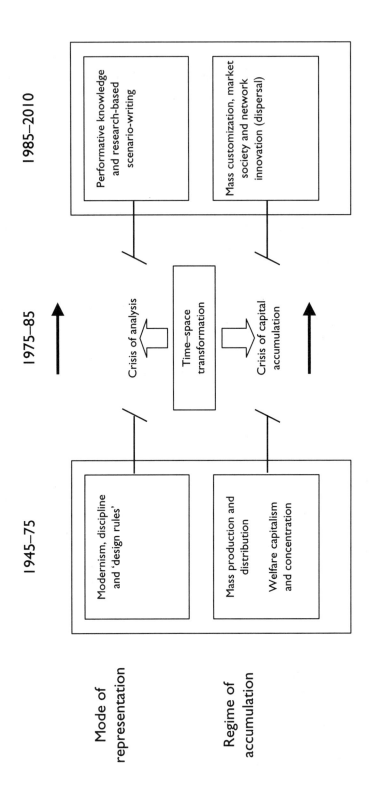

FIGURE 2.1 New political economy and mass customization (modified from: Gregory 1994, Figure 31)

costs of operation while increasing their revenue through reducing the turnover time of capital and reducing inventories. New possibilities in software were portrayed by the active suppliers of organizational innovations as being a network nirvana. Moreover, new spaces were aestheticized in respect of themes in the style of Disney. The new times at work were introduced by removing many of the co-ordination processes that slowed the turnover of capital. Most of these changes were concentrated in the USA, even when the source of new practices was elsewhere (e.g. Toyota in Japan).

Since 1985 the mode of the representation of knowledge has become more performative and the key knowledge-making role has moved out of the universities into the transnational professional and global firms. They are major players in the transformation. The earlier preference for simple design rules applied homogeneously in largely bureaucratic firms have been displaced by more modulated rules. Moreover, the uncertainties about the future are increasingly addressed by developing multiple scenarios (Hodgson 1999). Mass customization has become the rhetoric through which the faster pace of innovation in services, processes and products is explained. This hypothesis is useful in emphasizing the different role of the consumer in the market society (see Chapter 6). Design cycles are shorter and cosmetic changes more frequent. Advertising seeks to enrol the consumer in revenue-generating activities (e.g. fitness, beauty and serious leisure). The new information and communication technologies enable some forms of capitalism to innovate continuously in their organization of networks. As a result capitalism becomes more dispersed geographically while being more connected through global commodity chains (see Chapter 10).

Performative knowledge is distinguished by the ideal that knowledge should be actionable and commercially useful. Now it is claimed that even performative knowledge has a half-life of only seven years. This means that existing knowledge should be constantly renewed and useless knowledge – and the bodies it occupies – should be exnovated (taken out) (Clark and Staunton 1989). The corporations and professions of business present a discourse in which Schumpeter's notion of the creative destruction of old practices becomes legitimate and desirable. So, innovation should be continuous rather than exceptional.

Figure 2.1 stylizes many elements and therefore some indeterminacy should be introduced about Harvey's account of compressed, commodified time–space. Performance in this current era is now more *modulated*. This means that the notion of disciplinary society proposed by Foucault requires revision. Disciplinary societies replaced and displaced the societies of sovereign power during the Napoleonic era and reached their peak in late modernity (*c*. 1945–80). In disciplinary societies there were vast spaces of enclosure so that the individual was constantly moving from one enclosed zone to another. Thus the factory, the school and the hospital concentrated and distributed populations through time–space. Disciplines were defined and built in particular sites and their ideals were pursued to organize collective life within walls and rules. Within those special sites their members sought to discipline their bodies and identities to accomplish pre-designated time–space flows. Capitalism provided the relational config-uration that shaped the disciplinary enclosures. Enclosures became moulds producing distinct shapings of their inhabitants, both human and non-human. Thus nineteenth-century capitalism concentrated both production and property.

Now the previously enclosed and regulated spaces are being opened up. Now controls are deforming into modalities. Control is more free-floating and some new spaces are paradoxically liberating and enslaving (see Harvey 1989). The corporation is replacing the factory enclosure and its form of reward applies the modulating principle of salary according to merit rather than position in the bureaucratic hierarchy. Whereas pre-viously in the disciplinary society the individual possessed a number and a signature, in the modulated society the new code is password and individuals become part of data banks. Because capitalism is so much more involved in selling, it is dispersive rather than concentrated. Marketing becomes the soul of the firm.

In the modulated performance and control society its members gain new freedoms from their dispersal while being pressured by the owners of capital and their savants to engage in modulated performances. Time–space movement is entrained to the contingent variations of the work context such as unexpected seasonal oscillations (P.A. Clark 1985). Unless these contingencies are handled in a modulated way, capital loses profits and increases risks. Time–space movement now requires passwords ruled by precepts. Introducing the password means encountering the complex (hidden) architecture of software programmes, the electronic embrace, which sorts individuals into particular trajectories. Business colonizes everywhere with these new time–space requirements. The effects of modu-lation can be detected in the dress codes imposed by the large accountancy firms.

Harvey applies an historical-materialist analysis to claim that there has been a sea-change in cultural and political-economic practices since the mid-1970s which is linked to the emergence of new dominant ways in which time and space are experienced. These changes arose *during* the shift from Fordism to more flexible accumulation. The changes introduced a radical commodification and compression of time and space. These had a disorienting and disruptive impact upon political and economic practices, including the balance of class power and socio-cultural life. Harvey, Appadurai, Jameson and many of the post-moderns contend that this new volatile, chaotic experience of production and consumption is harrowing and so long-term planning becomes difficult.

Several trajectories unfold concomitantly in this transformation. First, the compression of time and space has been accomplished through rapid changes in forms of organization, in new techniques of production, and in the speed-up of the entire production process. New information tech-nologies and electronic networks play an important role. They accompany and enable an increase in pace and vertical disintegration towards sub-contracting and outsourcing. The structure of the global workforce is progressively transformed. Organizational innovation includes processes such as the Just-in-Time system, which speeds up the labour process by requiring a faster turnover time in the learning of skills to meet new demands. Second, there is a marked increase in the speed of financial exchange. The new electronic networks promote increasingly rapid trans-actions, such as electronic banking, which has enabled the flow of money to increase. Third, in consumption, new packaging and inventory control allow products to move through the market at greater speeds, which in turn depends upon the intensification of consumption patterns. This acceleration shortens the life-span of consumer products and fashions, production techniques and labour processes (Pine 1993) and increases the role and fusion of innovation with design (Clark 2000). In the emerging everyday life pre-existing values and practices are eroded and corroded (Sennett 1998). Fourth, Torres (2000) observes that faced by this volatility market actors engage in the manipulation of taste and opinion through the construction of new signs and images, propagated by the media and advertising. These signs/images are used to create new consumer needs and to structure consumption choices. Moreover, needs and choices have increasingly evolved to become commodities in themselves. There is a currency of images. Hence the former problems of volatility might be experienced, because of hegemonic structuring, as benefits when these images can, through increasingly powerful electronic technologies, be quickly changed and reproduced across space. Because of the increasingly prominent role of the media and information technologies, images and signs come to function as symbols of authority and power and are able to link a product with a certain social identity: branding and reputation (Schultz et al. 2000).

Three competing research programmes: their discursive formations

This section uses the concept of research programmes and discursive formations to express theories as structural wholes in which the meaning of concepts depends on the theory (Lakatos 1978). My usage of research programmes will be more influenced by the notion of multiple, com-peting relevant groups within a research programme than is Lakatos. We are theory laden (Archer 1995; Sayer 1992, 2000). The research pro-gramme is a structure that guides future research by stipulating the under-lying background assumptions (Searle 1995). These taken-for-granted assumptions are the hard core of the theory and are saved from falsification by a protective belt of auxiliary hypotheses. Theories are sometimes at the leading edge (progressive programmes) and then become lagging (degenerating).

We identify three different, co-existing and influential programmes through which the field of innovation is theorized. These three research programmes are competing and the early research programme certainly retains exponents.

The field of innovation is diverse and that diversity is both a problem and a source of rich experience in practice and research. The notion of research programme is used to refer to a community-like collection represented by their definition of sound knowledge and the critical problems.[1] Typically, research programmes contain several major and many, many minor research programmes. There tends to be a dominant discourse surrounded by a protective belt of fuzzy propositions hidden in background assumptions. New emergent research programmes tend to be multi-centred and still evolving. Each research programme shares some elements that might be used by 'historians of the present' stylistically to locate persistence, dislocation and new clusterings. However, even a dominant cluster may be more precarious than apparent, yet possess limited degrees of freedom. Once established, research programmes tend to retain their founding cohort whose influence pervades later periods beyond the era of founding.

Foucault (1972) demonstrated that institutions wield power and domination in particular epochs because a *discursive formation* becomes sedimented in both practices and conversations over epochs. This is an extension of hegemony. In every institution discourse is inseparable from power while being the medium of governing and ordering events. Discursive formations govern what it is possible to say and think. The practices associated with particular forms of discourse consist of sets of rules and procedures governing thinking, writing, and even performing. Individuals working within a particular discursive practice obey the unspoken archives of rules and constraints. Therefore, revealing the discursive formations governing the present is almost impossible. However, it is possible to go backwards (archaeology) to find transition moments from one formation to another and then to come forwards towards the present (genealogy). Discourses are produced in power struggles and represent the sedimented outcomes, including the truces. A discourse and practice typify long epochs of generations and centuries. History is therefore a discontinuous succession of discursive practices. Foucault made the analytic choice to abstract discursive formations from ongoing power struggles and to situate them in national cultural repertoires. In that respect they illuminate the more simplistic usage of mentalities in Braudel's historicizing of world economies. Calendrical history is therefore transformed into a discontinuous range of discursive practices. The fundamental discursive shifts cannot be matched against precise time because Foucault does not explore them in conventional historical time.

---

1   My usage of Lakatos (1978) is more general and loosely textured but with close reference to the relevant groups within a major research programme.

Foucault's (1977) analytic strategy demonstrated that wer was the site where forms of governance shifted in the eight from use by the sovereign of public punishments, such as ha and quartering, as a formative discourse to their replace' punishments. Compliance in the nineteenth century w through external fear and more through the internalizatiu.. disciplines and bodily modalities: the disciplinary society.

It has been a short step from Foucault's *œuvre* on the discursive formations that unfold over centuries in western Europe to the construction of discursive formations in organizations everywhere. Discourse now has an analytic role in explaining the seeming epiphenomena of small-scale, shortterm changes in a single strata of an organization. It seems that for some academic analysts the mission statements of take-overs become, almost overnight, the new managerial discourse. That usage both undermines Foucault's contribution and also underplays the relevance of the new institutionalism (see Chapter 5).

The apparent dominance of a research programme eventually encounters an array of critiques. These lead to the eventual loosening of the protective belt of fuzzy assumptions. In this period of challenge and of the emergence of a new research programme, the new critical problems are highly salient and are sometimes expressed by using 'strawmen' to exaggerate the limits not addressed by the new critical problems. The incoming research programme tends to contain critical problems whose earlier presence and future salience were not apparent. For example, the sociology of science in the 1960s was radicalized in the mid-1980s by the social construction of technological systems and heterogeneous engineering (e.g. Bijker, Hughes, Pinch, Callon, Latour, Woolgar and Law).

Research programmes compete for control of particular fields. Consequently, several research programmes may be in play at the same time. This has been the position over the last decade and is likely to continue for much of this decade. For convenience we can identify three major research programmes over the past four decades. In order of founding, these are:

(1) Innovation as technological artifact (1950s–1980s);
(2) Innovation as technology-and-organization (1980s–1990s);
(3) Network innovations (1980s onwards) in 'electronic embraces' (1995 onwards).

The earlier influence of 'technology' is caricatured in the second and third programmes as 'technological determinism'.

The modern programme: innovation as technological artifact

The research programme of 'innovation-as-technological artifact' gained salience in the 1960s, remained dominant into the 1980s and retains exponents today. This research programme was strongly influenced by economists specializing in the national policies for investment in science and technology. Consequently, the social and organizational aspects of innovation were depicted as contextual and therefore secondary to the material artifacts. In this early period many social scientists shared the basic assumption of technology as an artifact to which organizational systems should adjust. In this research programme the definition of problems objectifies innovations and equates innovations with material artifacts such as saw-milling machines, steam engines and new genres of aircraft (e.g. Concorde, 747s). Analytic attention is given to showing how material artifacts contained new combinations of raw materials and/or new equipment in process and product. This is illustrated by the development and design of the tufted carpet at the Georgia Institute of Technology in the USA and its diffusion into Europe (Clark and Probert 1989). There is a tendency to chart diffusion patterns rather than explain its presence in one location (e.g. America) and its absence in another location (e.g. Italy). Only a small number of organizational artifacts were researched. These research programmes shared a tendency towards the modern and positivistic vision of scientific enquiry and therefore aimed to compact knowledge into law-like propositions from which scientific experts would be able to legislate to potential users and to the general public about what ought to be done. Consequently, alternative explanations that focused on organization rather than technology were ignored and suppressed. There are two intriguing examples of compression, evisceration and suppression in the seminal studies by Burns and Stalker (1961) of the management of innovation and in Chandler's (1962) historical study of organizational innovation. These will be discussed later.

Five sub-programmes were aligned to the technological artifact research programme:

(1) The *left–right linear models* of how science was invented in the universities (on the left) and became transformed into product and process innovations as technologies by firms. This theme was dominated by notions of 'Big Science' undertaken within the major universities and specialist scientific centres (e.g. MIT) from where – somehow – abstract, codified knowledge was transferred to firms for exploitation. Economists were major contributors to policy at the national level.

(2) The linear models of the *diffusion of innovations* proposed by Hagerstrand and by Rogers. These diffusion theories arose in economic geography and in the socio-psychological communication theories. They soon became part of the knowledge used by marketing departments and in early consumer research.

(3) *Overcoming resistance to change* through the participative educational frameworks of planned organizational change (Bennis et al. 1961; Lupton 1964). Resistance was presumed to be irrational and

anti-modern. It was assumed that the Lewinian three-step model provided experts with usable knowledge. Failures were ignored.

(4) The *contingency theory of innovation*. The distinction between mechanistic and organic management systems was used to highlight environmental uncertainty while reducing the problematic of organizing into a simple 'can do' recipe (Zaltman and Duncan 1977).

(5) *Socio-technical systems and Quality of Working Life (QWL) models* of organizational choice proposed by the Tavistock group (e.g. Trist, Emery and Miller) were frequently mentioned.

Although much of the socio-technical school fitted rather awkwardly into this core programme, its participative rhetoric and the autonomous group were selected. Other elements were discarded. The approach to organization design of Miller and Rice was largely ignored (see Clark 1972, 1975, 2000).

There were also three very different seminal texts which provided the basic frameworks for exploring the organizational aspects:

(1) *Rogers* (1962), on the diffusion of innovations by suppliers to users inside organizations of all kinds (Rogers and Rogers 1976). In the case of Rogers' book its whole format was absorbed, compressed and developed. Rogers (1962) had defined and set the scope of the study of the diffusion of innovations and it had become not only the most highly cited text in the field, but one of the most highly cited texts in all the social sciences. His small, highly readable handbook explained how the supplier should market an innovation in order to achieve a sale.

(2) *Burns and Stalker* (1961), on the internal management of innovation within the firm. Only a very small part of their analysis was included (1961: 119–125) and much of their contribution was eviscerated and neglected (Clark 2000). Even the famous SAPPHO programme led by Chris Freeman at the Science Policy Research Unit (SPRU) reconfigured this text.

(3) *Chandler* (1962). This 400-page book was compacted into a few paragraphs. The path-breaking analysis of the multi-divisional form as organizational artifacts by Chandler (1962) was misunderstood. Indeed, even Chandler may have underestimated his own contribution at the time and this is acknowledged in his 1989 Preface to the third edition of *Strategy and Structure*. Chandler's work illustrates a research programme from an historical perspective that remained largely attenuated until the much later study of electronic power networks as seamless elements in the social construction of technological systems by Hughes (1987).

The influence of these three continued for four decades but their interpretation altered within each research programme.

The limits of this research programme were signalled and prised open by three new research programmes in which organization was of equal importance to technological artifacts. First, Freeman's development of Schumpeterian long waves. This had a distinct impact. Second, Abernathy's (1978) use of organization theory to formulate the innovation/efficiency dilemma. Third, the publication of the second edition of the *Diffusion of Innovations* by Rogers in 1983 revised core assumptions in the earlier publication. Attacking the pro-innovation bias (see Chapter 7) involved a major shift in background assumptions.

In this research programme and during its period of ascendancy there was a widespread assumption that corporations could pursue efficiency with short interruptions by exceptional periods of innovation. A major feature was incremental innovation through the experience curve. Abernathy (1978) pointed out that the major dilemma centred upon how to be efficient. This stress on efficiency rather than innovation enabled the suppliers of innovations (e.g. IBM) to develop the capacity to orchestrate the innovation process and to choreograph the tastes of their consumers.

The incoming ontology took 'technological determinism' as the strawman.

Anti-modern programme: duality of technology-and-organization

This period (1980s–1990s) was enlivened with the regular and ritual defeat of the strawman of technological determinism. However, the negative case was insufficient. The critique of technological determinism was rampant but was too often used rhetorically. Even so, the protective belt of fuzzy assumptions around innovation as a technology artifact was exposed and the organizational aspect gradually became more equal in its attention.

Developing the duality programme required innovation being re-conceptualized as 'technology-and-organization'. There were a growing number of studies that gave equal attention to the *duality* of:

- technology as hardware and artifact
- social organization as artifact.

The difference from the earlier period is the growing equivalence implied by the 'and' so that it became 'AND'. Several studies used the small 'and'. For example, the careful empirical study of information technology in three sectors and six European nations by Child and Loveridge (1991) exemplifies attention to technology as hardware, as knowledge and as principles of organizing. Yet these are juxtaposed. The design of that project was too early to benefit from either the full theoretical insight of the structuration and social constructionist theorizing or from the possibilities inherent in the software developed later for spread-sheet comparisons (see Stymne 1996).

In the 1980s the recognition of the social and organizational elements increased markedly.

Those who insightfully sought to embrace and address the potentials in the structuration and agency debate attributed to Giddens occupied another important transitional development. The structuration perspective influenced British studies of innovation in geography (e.g. Gregory 1982, 1994). One of the few organizational studies combined a variant of structuration with the sectoral perspective developed from Abernathy (1978) and by Whipp and Clark (1986). There was a much more direct engagement with structuration by American organization theorists and by those studying various innovations in information technologies. The most significant studies were by Barley (1986), using structuration and the triggering effects of new technology, and by Orlikowski (1992, 2000) and the notion of duality between organization and technology. All these studies share a common concern to re-theorize technology away from its treatment only as hardware (e.g. Orlikowski 2000).

Equally important was the emergence and growth of perspectives that are agnostic about the differences between human and non-human systems (e.g. actor network theories).

There were major developments in technology that linked control to the transformation and transport elements developed in previous technologies. These confirmed the relevance of Chandler's (1977) contention that transport and communication technologies are the critical problems in capital accumulation. At this point it is useful to adopt Bell's (1978) three-fold distinction of artifactual technologies as:

(1) *Transforming machines* that alter the state of a product or assembly into a new state. These were developed first, for example, Blanchard's machine (*c.* 1820) for shaping the stock of a rifle at Springfield, Massachusetts. Another key example was the Bonsack cigarette maker patented in the 1860s and commercialized in the 1880s (Chandler 1977).

(2) *Transferring machines* that carry the product/material from one phase to another phase. These became linked with the transforming machines. For example, in the production of automobile engines (Abernathy 1978).

(3) *Control machines* that were initially inserted into the transforming and transfer machines. From the 1960s onwards control machines connected the transforming and transfer machines into a hybrid network. An early example was the sugar beet factory (Clark 1975). In the 1960s control machines provided an initially utopian image of the future (Clark 1972), but the development of the Internet (*c.* 1994) actualized the electronic embrace.

In the late 1960s cigarette factories, which were at the front of technological innovation since Bonsack, were able to reduce their throughput time from 100 hours to well under ten hours (Clark 1972). In the next generation developments in computer software and hardware transformed information processing and were used to provide multi-module packages such as Materials Resource Planning (MRP, MRP2). These were further developed into the strategic technologies of Enterprise Resource Systems.

There were revisions to the previous clustering coupled with a growing critique of the dominant cluster. The revisions included:

(1) The left–right model became the focus of scepticism by, for example, the explorations using the notion of *knowledge chaining* as an interactive rather than linear framework (Kline and Rosenberg 1986). Moreover, long wave theories highlighted the possibility that periods of innovation might vary systematically in the intensity of innovation over long-term periods such as 40 or 50 years.

(2) The diffusion of innovations perspectives continued to be influential, yet were shown to possess a *bias in favour of the supplier* (Rogers and Rogers 1976) and to encourage a pro-innovation mentality (e.g. Rogers 1983).

(3) The resistance to change theme was absorbed fully into the disciplinary practices of Organization Development and Change with *a whole panoply of change technologies commodified and distributed through consultancies* (Wilson 1992).

(4) In order to cope with efficiency and innovation the mechanistic/organic solution was transposed into the notion of *the ambidextrous organization*.

(5) *Organizational capabilities* were increasingly theorized in the resource-based theory of the firm (e.g. Grant 1998; see also Clark 2000).

In this research programme four critical issues emerged:

(1) There was a growing interest in the probability that *national social structures constituted key building blocks and barriers to an innovation* capability. Nations differed in their capacity to generate and absorb major innovations (Nelson 1991). This critical problem was stimulated by an early intrigue about Japanese success. The comparison between the production of television sets in Japan and Britain (Dore 1973) was given extra credence by the closure of the British factories and the import of Japanese products followed by the opening of Japanese transplants. In the USA the growing concern about Japan's penetration of the profitable sectors (e.g. automobiles and consumer electronics) was analysed and revealed, especially by Abernathy and Hayes (1980). There was increasing attention to international comparisons and there were references to 'national systems of innovation' (McMillan 1985; Clark 1987). Much attention focused upon characterizing the elements of Japanese success. In the USA there was a switch from the comfort of Anglo-American comparisons of economic performance to the discomfort of Japan–USA

comparisons. There were moments of real despair for some American analysts. Rosenberg and Steinmuller (1988) were sceptical of the American capacity to adapt to the future and this notion was evident in the over-stretch thesis of Kennedy (1987). Abernathy, Clark and Kantrow (1983) were more analytical about the future.

(2) Research on the *total innovation-design process* from design concept through its translation into the commissioning of the new organization and then into the utilization revealed considerable re-working of the initial concepts (Clark 1987; Clark and Starkey 1988). Some studies illustrated how those to be affected by the innovation might engage in pro-active resistance (Whipp and Clark 1986). There was a range of examples in which the users transform the innovation. The simplest example is where banks introduced charge cards in order to earn interest, but a marked proportion of the customers simply used the charge card as a record-keeping device and did not borrow. The more complex example is in the use made of system-wide software packages that are designed to monitor and co-ordinate resources (e.g. Enterprise Resource Planning).

(3) The *social construction of technology and actor network theory.* This approach inserted a relativist challenge to science. This programme seemed to dissolve the differences between humans and non-humans while also highlighting the processes through which new systems are socially constructed. The seminal example is Hughes' analysis of electric power systems in the USA. Hughes examined the competition between the two main forms, DC and AC, and then compared American urban systems (Chicago and New York) with London to show how, even with a scientific and technical system, there was both social construction and social effects. The publication of *The Social Construction of Technology* (Bijker et al. 1987) publicly launched the key perspective, introduced key concepts (e.g. actor network) and provided a rich diet of varied case studies.

(4) Abernathy (1978) observed that the dilemma for firms now centred on more or less *continuous innovation* rather than periodic innovation.

These programmes provided the leading edges in the next period. They also deployed a new discourse in which 'network innovations' and *systemic innovations* were increasingly mentioned.

New political economy programme: organizational innovation and hybrid networks

The recent and current period radically differs from the earlier two periods in the content of leading-edge theorizing and also in the roles attributed to innovation.

The pro-innovation bias tended to treat the choosing of innovations as an efficient choice model applied by top decision-makers in rationally designed organizations. The process of innovation was treated as an abstraction of time-and-space (Clark and Staunton 1989). From the early 1980s onwards the claims of the pro-innovation perspectives were increasingly challenged, especially by the social constructionist cluster of perspectives (e.g. Mackenzie, Latour, Law, Woolgar, Callon, Bijker, Hughes) and by the new institutional school in organizational sociology (e.g. DiMaggio, Powell, Scott, Fligstein) who proclaimed inefficient choice (e.g. Abrahamson). To an increasing degree analysts perused the singularities and specificities of event and place with a sharp increase in quasi-historical descriptions of innovations in firms and national contexts (e.g. Tidd and Fujimoto 1994). Innovations were viewed less and less as clear templates to be applied by a knowing community of expert practitioners inside organized boundaries and more and more as a hybrid collection of possibilities that needed crafting to the contingent dynamics of an emergent situation. So, the explicit knowledge emphasized in the pro-modern perspectives was shown to be defective in its account of the role of tacit knowledge. After a tidal wave of articles on tacit knowledge and the rediscovery of the routine and habits, attention turned to the blending of tacit and explicit – explacit knowledge – and the issue of organizational repertoires and recurrent action programmes (Clark 2000).

After the mid-1980s some multinational corporations were transformed from a vertically integrated, multi-divisional bureaucracy into a networked business in which design, or innovation-design, became the central value-adding capacity. The focus upon design recast the role of research and of research and development (Clark and Staunton 1989). Consequently, the left–right model of science into innovation has been transformed into an interactive model linking the end-user and basic science through chains, networks, circuits and webs (e.g. Kline and Rosenberg 1986). Also innovation was more concerned with the processes of computer-based control than with the processes of transformation and transportation.

Over the past two decades there have been major transitions in the analyses of innovation. These changes in perspective have accompanied the expansion of the domains to include and highlight national predispositions. Six critical problems are frequently cited:

(1) Conceptualizing and explaining the domain of *national systems of innovation and the problems of international transfers of innovation.* The national systems of innovation literature was developed initially by industrial economists (e.g. Freeman, Nelson, Lundvall, Dosi, Pavitt, Dicken) and then by social scientists (e.g. Whitley, Sorge, Child, Clark). There were four main forms.

   • Exploratory analysis of a single nation. Pavitt (1980) insightfully examined the British case and was among the early exponents of the notion that contexts are formative to the success/failure of domestic firms in their international affairs. A plethora of studies sought to characterize innovation in Japan in order to provide templates and benchmarks of best practice to be applied elsewhere (e.g. Freeman).

- A plethora of edited collections of nations. Dosi (1984) examined the evolutionary trajectory of information technology to highlight differences between the USA and Europe. Nelson (1993) edited a collection of authors, each characterizing his or her own national systems of innovation against a loosely defined framework.
- Porter (1990) developed a new framework (see Chapter 5).
- Awareness of the international issue led the European Commission to invest in large-scale research including the European Innovation Monitoring System (EIMS).

Gradually this research began to refute and revise the pro-innovation perspectives. Fujimoto (1994) revised earlier claims about the Japan-ization of innovations in Toyota by detailing long-term American influences in the 1930s and 1950s. His account confirmed blending and hybridization. Research on the British automobile industry also revealed the firm-specific and national sectoral features that precluded American and Japanese emulation, while failing to protect the domestically owned industry against foreign direct investment except for sports car racing. Studies of the transfer of systemic innovations between nations drew attention to:

- to the difficulties of transfer (e.g. Jeremy 1981; Clark 1987, 1997)
- to the extent of differences between originator nation and the emulator nation. Abo (1994) devised a set of scales to benchmark the adoption of Japanese innovations, thereby confirming large-scale hybridization.

The case for expecting non-transfer was starkly put by Clark (1987) with respect to innovations travelling between the USA and Britain. Many of these studies centred on the capacities of American firms selectively to ingest key features of Japanese innovations in the usage of space and time and in the application of high standards of quality. There was slow application of these revisions to the dominant diffusion of innovation theory.

(2) *Institutional theory* – the roots of new institutional theory lie in the critique of neo-rational educational reform in America by Meyer and Rowan (1977) and were crystallized in the seminal article by DiMaggio and Powell (1983). Scott (1995) summarizes the connected array of developments.

(3) *The inefficient choice thesis and the role of fashions.* There are many tributaries to the inefficient choice thesis (e.g. Abrahamson, DiMaggio and Powell, W.B. Arthur and P.A. David). Role of fads and fashions is examined Chapter 7.

(4) The *social construction* movement, with its strong Anglo-French hub, has generated much of the new conversation about innovation. The early studies of scientific activity – mainly in French laboratories and scientific history (e.g. Pasteurization in France) – gave renewed impetus to the social construction of knowledge and to the thesis that technological artifacts are social artifacts 'made durable'. Not surprisingly, it is the social artifacts that seem the more durable (as in Gurvitch 1964).

(5) *Hybrid network innovations and the electronic embrace.* The strategic co-ordination of firms and between firms is an area most likely to be impacted by national predispositions and national sectoral clusters. If so, the growing array of organizational innovations, blended with the hardware and software of information technologies, is a key area for investigation with regard to both national performance and also corporate performance. Organizational innovations move to centre stage and their unfolding alters the remit of orthodox organization studies (Giddens 1990, 1998). The contribution of Castells (1996) is of major importance (see later).

(6) The critical problem of *what is knowledge* is being prised open through the confluence of a large cluster of fresh perspectives:

- Performativity of knowledge (e.g. Lyotard)
- Problematizing of 'grand narratives' such as Marxist theories
- New Production of Knowledge (e.g. Gibbons et al. 1994; Huff 2000)
- Power-knowledge regimes (e.g. Foucault)
- Communities of practice (e.g. Suchman, Brown, Lave, Wenger, Cook and Duguid)
- Enacted socially constructed environments and variation-selection models (Weick)
- Activity theory (Engestrom, Scribner)
- Creating knowledge (Nonaka) and knowledge assets (Boisot).

We shall examine these further.

Hybrid networks: Castells

Setting the scene

According to Castells (1989, 1996, 2001), information technologies are not a tool, but are organizational processes with high interpretative flexibility that are shaped by both suppliers and users. Informational capitalism has replaced industrial capitalism. The network society is the new form of informational capitalism and is the outcome of a conjuncture of three autonomous processes:

(1) Information technology revolution.
(2) Resolution of certain contradictions in industrial capitalism that became evident after the fiscal crisis of the mid-1970s.
(3) Various cultural movements, including feminism.

These three autonomous processes were strongly driven from the USA, possibly with a West Coast edge. Their conjuncture enabled the American confrontation with the Soviet Union and the transformation of the Cold War through the emergence of the Pacific Rim. Therefore, the evolution of information and telecommunications technology since the mid-1970s has been firmly anchored in the Cold War policies of the USA. Moreover, within the USA there have been multiple zones of manoeuvre for American firms and for the highly agentic new service class. Castells comments that technology does not determine society: it embodies it. But neither does society determine technological innovation: its uses it (Castells 1996).

Castells' analytic style is to *retroduce* a geo-history of small narratives in the new political economy genre and to show both intended and unintended, emergent outcomes. This is a relational materialist theory with considerable potential. Putting those stories together in an analytically structured narrative of many cameo studies is readable and revealing. Castells shows how in the use of the new process, organizational innovations were co-shaped by the users. The new information technologies have transformed the knowledge-making that surrounds the price mechanism, enabling hierarchies to be marketized and to transform the market for business-to-business transactions.

Castells is preoccupied by the role of events in the USA. His analysis provides a stark counterpoint to the claims of other nations to be central players, especially the United Kingdom. It seems that Britain entered the twentieth century with the wrong competitive context for the long-term survival of its domestic automobile industry and ended the century with a subordinate and peripheral role in the new modes of information processing.

In the west a new informational mode of techno-organizational innovation was established in the 1980s and 1990s (Castells 1989: 13). The techno-organizational revolution has transformed and is further transforming our fundamental understanding of space and time, but not quite in the ways posited by Harvey (1989). A key factor in the process is the massively enhanced ability to store, retrieve and analyse information. Process-oriented organizational innovations can penetrate all spheres of administrative activity at work and in the home. However, the productive forces within a society are only enhanced when there is a capacity to educate and motivate the labour forces, when there is an institutional structure that maximizes information flows and connects them to the development of core activities: 'The more a society facilitates the exchange of information flows, and decentralized generation and distribution of information, the greater will be its collective symbolic capacity' (Castells 1989: 15).

From the 1930s to the 1970s western societies relied on the neo-corporatist pacts composed in the welfare state to steer a complex society. Western capitalism relied on three structural modifications (see Figure 2.1):

- a social pact between capital and labour
- regulation and intervention by the state
- control of the international economic order.

That period was often referred to as the golden age of western capitalism.

After the crisis in capitalism's rate of profitability in the 1980s, the role of firms and of the state became transformed in the USA and in European economies. A crucial part of that transformation centred upon the new capabilities for information processing and knowledge creation. These new capacities strongly contributed to the isolation and collapse of the communist bloc, especially the USSR. A new model of socio-economic organization emerged in the 1980s with capital appropriating a higher share of the surplus from the production process through higher productivity, lower wages and decentralization to low-cost regions and through transferring functions from the welfare state. Labour markets were restructured and the trade unions were weakened. The role of the state became one of political domination and capital accumulation rather than political legitimacy and social redistribution. Many state activities were deregulated and privatized. Regressive tax reforms were introduced. The state gave priority to defence-related industries, especially in the USA, and to the support of R&D. Fiscal austerity increased. Economic processes have been and are being internationalized to increase profitability through the expansion of the system. This enhances profitability by allowing finance capital to search for the highest returns from anywhere.

Information capitalism

In information capitalism there are *two fundamental modalities*:

- the focus upon information processing
- the main effects are on processes and organization rather than products.

These *articulate* in different ways with different consequences. The notion of articulation and of articulation errors is subtle, complex and central.

First, the new modality of informational technology can enable three processes:

(1) An increasing rate of profit through micro-electronic equipment, coupled with decentralized production. So, management have and are able to automate many assembly and transfer tasks and labour's zone of manoeuvre is confined by capital.

(2) The new technologies give the state more zones of manoeuvre for domination and accumulation.

(3) The new technologies enable the time–space stretching to deepen the internationalization of the economy.

Second, in the organizational modality there is:

(1) A growing concentration upon knowledge-generation and decision-making processes in the upper levels of firms in the core centres. Former corporate bureaucracies lose their political and integrative functions to the new technocracy and new managerialism. Knowledge holders become increasingly powerful.

(2) Flexibility is enhanced through new, easy-to-meter contracts. Flexibility enables the state to engage in public–private contracts. Capitalist firms are better able to adapt to changing conditions.

(3) Firms shifted to hubs and decentralized networks containing a diversity of sizes and forms of organizational units.

The articulation of these two modalities varied between different contexts in the world economy, thereby affecting the location of seedbed milieux for high-technology industries in different states. For example, so far, the USA contained milieux that occupied major positions and the UK contained no significantly large milieux.

Because of these developments it is necessary to distinguish between the high-technology producers and high-technology users (Castells 1989: 35) and their different social construction of spatial locations (ibid. 1989: 75; Castells 1996). First, the large firms that dominated semi-conductor production constituted isolated, self-sufficient production systems as in IBM, Texas Instruments and Motorola. For example, the locational pattern of IBM was influenced by the capacity of the firm to control the high rates of obsolescence of the new technology through vertical integration in the computer market. IBM did not rely on proximity to research centres, preferring to endogenize, and developed an ethos of isolation with campus-like facilities. Likewise Texas Instruments and Motorola remained independent of specific locations. Moreover no single company can generate an industrial milieu. Thus the heart of the semi-conductor industry occupied varied locations, including Silicon Valley and Boston's Route 128. These two locations have core features in common: an exceptional concentration of research-oriented university complexes; an active and organized network of financial firms; positions as major regional centres and nodes of international telecommunication and air transportation. A crucial characteristic is access to the military markets whose role was so crucial in the 1980s. In the case of the semi-conductor industry, their specific labour requirements were associated with low unionization. Finally, the industry possesses a sharp spatial distinction, integrated circuit manufacturing establishments and discrete device manufacturing (Castells 1989: 54). Second, the computer industry is larger and more diversified because there is a key linkage between production and the use of the technology. For example, the original reason for IBM's location in New York was that the machines were so expensive and were leased to firms which, because they lacked computer knowledge, needed direct access to maintenance, repair and services. The computer industry decentralized, as with IBM setting up in Europe. The computer industry 'sells pure knowledge, fabrication being reduced to the minimum material expression' (ibid. 1989: 66). Even so, there was concentration in California because their key employees preferred specific life-styles. Moreover, there were major research-oriented hospitals and health institutions working on government-sponsored contracts whose usage of the technology provided learning by using to the suppliers. There are fundamental characteristics of the computer industry whose specific connections also require examination. These industries rely heavily upon high-quality information from scientists and technical labour. The production of process-oriented devices requires the capacity to adapt new products to the needs and desires of users. The combined effect of these tendencies explains the very distinct internal segment of the production process. These industries are pioneers in using their own products in the production process. Therefore four spatial processes may be noted:

- sharp spatial division of labour
- information generation dominates
- production functions are decentralized with each location producing the original template
- high flexibility.

Castells contends that our conventional notion of the service sector is flawed.

Now organizational systems of corporate co-ordination involve the convergence of two streams: the computerization of information processing and the application of telecommunications to information exchange. First, corporate and state organizations went through three steps:

- mainframes building databases
- micro-processors controlled by users without the intermediation of computer experts
- heavy reliance upon networking and integrated office systems enabled by telecommunications.

The dramatic expansion of transmission capacity gave birth to information systems. In the latest stage routine operations were automated and the process was rationalized. This altered the dynamics of administration by increasing the economies of scale and scope, increasing the complexity contained in outputs, and by interconnecting distributed activities. In the firm, its durable structural features became transparent and were increasingly analysed as flows and processes. The space contains flows. There is an expansion of information-intensive industries and their location in metropolitan areas where these processes provided local area networks. Office work could now be centralized and its routine operation could be controlled, co-ordinated and metered as in the theory of transaction costs. The process is differentiated with routine tasks following the hierarchical and functional logic of teamworking and new steering centres doing the knowledge-intensive work. There is a dominance of the large organization with decentralized management and the subcontracted outsourcing of many operations. The firm becomes a network located within other networks. Thus major innovations in information technology have provided an agentic opportunity to alter rather than entrench existing knowledge and occupational structures. The specific situation depends on the specific relationship between the two processes (Castells 1989: 169). Castells concludes that the space of the firm becomes the space of flows and therefore firms can disembed themselves. However, it is important to acknowledge that the new technologies were used to develop the knowledge technology of system dynamics.

One of the consequences is that the space of the city becomes separated between opulence and poverty. Castells contends that the majority of new occupations do not require sophisticated skills and that there is a dismantling of the capital–labour relationships (Castells 1989: 224). The middle levels are hollowed out to create a spatial metropolitan space combining segregation, diversity and hierarchy. Thus in the 1980s the urban welfare state is transformed through an anti-welfare strategy and a pro-warfare strategy with a large build-up of a new power bloc at the core of the warfare state (ibid. 1989: 236). Business took the initiative in these developments. It was the military use of technology that became the driving force in the development and application of scientific discoveries (ibid. 1989: 262). This including the Strategic Defense Initiative and the billions spent with the scientific community (e.g. Stanford and the APARNET). The scientific lobby proffered the search for a military El Dorado. The American technology-driven defence policy had significant effects on both firms and the urban context.

Since the late 1970s there has been a growing integration of the US economy in international trade with far-reaching effects for the world economy and the USA (ibid. 1989: 312). In the mid-1980s there were deep negative trade balances with the world and that was defined as a crisis. At the time it seemed that American competitiveness had been eroded (ibid. 1989: 314), especially in the automobile industry (ibid. 1989: 320–331). Kennedy (1987), not mentioned by Castells, seemed to provide the required explanation of imminent decline.

The information technology revolution creates administrative processes that are pervasive. The application of knowledge and information to knowledge generation and information-processing devices is a cumulative loop between innovation and the uses of innovation. According to Mokyr (1990), technological revolutions take place in only a few societies and are diffused to relatively limited geographic spaces. There are local seedbeds.

Competitiveness has come to the foreground. The linkage from information technology and organizational change to productivity growth is largely mediated by global competition. Informational capitalism revealed a productivity potential in mature capitalism and led to the repoliticizing of the nation state. Now the global economy is more tightly linked, both for signal information on the changing value of firms and for the mobility of finance capital. The increasingly globalized economic agents are in a complex interaction with historically grounded political institutions. There is extensive regional segmentation and asymmetry of the global economy (Castells 1996: 100–101). Many areas are excluded (e.g. Africa). The ex-Soviet republics face segmented incorporation into the informational economy. Thus the structure of the economy combines an enduring architecture with a variable impact on regions leading to a new division of labour (Castells 1996: 147).

Network enterprise

In this transition from Fordist/welfare political economy (see Figure 2.1) there are new organizational trajectories replacing the vertically integrated firm:

(1) Large firms and medium-sized firms face a crisis of survival (e.g. Marks & Spencer).
(2) New methods of management have been identified, many originating in Japan. While American firms gained productivity increases from specialization and job demarcation, the Japanese firms engaged in what we now term 'knowledge creation'. New knowledge is related to innovation (e.g. Nonaka and Takeuchi 1995).
(3) There has been the development of extensive inter-firm networking. Toyota is a symbol of the network organization and also invests heavily in network building.
(4) There has been the development of strategic alliances specifying co-operative contracts.

Consequently the new millennial corporation is horizontal (Castells 1996: 164) and is organized around processes with team management. Performance is measured against customer satisfaction, both actual and anticipated future satisfaction. The notion of the lean thinking and the lean enterprise received considerable attention. Gradually, the actual operating unit becomes a business project enacted by and through a network (ibid. 1996: 165). The major problem to be faced is the gap between what the consumer wants and what the supplier makes available. In London the new Tate Modern museum of modern art gained customers while the Dome became the symbol of failure. The gap is elegantly and simplistically referred to as articulation errors. *Now networks are the fundamental texture of the new horizontal organization.*

Castells aims to define the network enterprise by acknowledging the distinction in evolutionary theory between two types of organizing (ibid. 1996: 171). One type – bureaucracies – has the goal of reproducing the system. The other type – network enterprises – has the goal of continually reshaping the structure and means. Castells defines the network enterprise as a specific form whose system of means is constituted by the intersection of segments of autonomous goals. The components are both autonomous and dependent on the network. There is connectedness. There is ample evidence from comparative studies of management that the western variant of the network enterprise is just one possibility (ibid. 1996: 190). Therefore, there is a search for a concept that symbolizes flows and connections through the use and consumption of material and symbolic bases of information technologies.

Firms and societies should be conceptualized as spatio-temporal flows: their social morphology. The information revolution is re-shaping the material base of these flows. This is leading into the reconfiguration of user-firms and of user-life-styles. The diffusion as a process is reinforcing. Castells (1989, 1996) contends that the new informational mode of development is altering the ways in which capitalism *embeds and disembeds the social morphology of society*, that is spatio-temporal flows, and these possess eight features:

(1) Social actors hold purposeful, repetitive, programmed sequences of action and exchange involving interaction between the spatially disjointed positions.
(2) Societies are constructed around flows. The dominant firms shape the dominant spatio-temporal rhythms and flows.
(3) Spatio-temporal flows articulate and represent the social morphology of work and society. They are the material organization of time-sharing social practices through which work flows (Castells 1996: 412f).
(4) Key dimensions are commodification, compression, *stretch and colonization*.
(5) The dominant spatio-temporal flows are networked, seemingly ahistorical and contain hidden organizing codes.
(6) The spatio-temporal flows possess three main layers:
  (a) The *material base* and all its artifacts.
  (b) There are hubs and nodes because these are defined according to their role in international spatio-temporal flows. Technological *infrastructure* expresses, defines and colonizes from its 'hubs'. The materiality of the spatio-temporal flows invades the domestic scene (e.g. Silicon Valley Culture Project).
  (c) The role of the *international elite of the service class* is in the politics of assembling more economically powerful knowledge chains through a variety of strategies (e.g. divide and rule between places as in Liverpool for Jaguars and Dagenham for the car boot). They possess the capacity to organize and ignore some

milieux and strata. Elites create and fund new life-styles in symbolically secluded communities so that there are hierarchies of eliteness. The actions of elites shape the space of places (e.g. Birmingham compared to San Jose) The new spatio-temporal flows require considerable, continuous interaction between all segments and parties, potentially including households.

(7) The spatio-temporal flows commodify activities and heighten the compression and dis-articulation of everyday life:

(a) The spatio-temporal flows of production have been compressed through integrating transformation and transfer technologies with control technologies. The full impacts of control are just unfolding.

(b) This alters the balance between Fordist templates of control time as linear, explicit, measured, located in temporal inventories and schedules like Enterprise Resource Planning Systems (ERPS) from the 'new times' (Clark 1997).

(c) Financial trading involves tightly articulated almost synchronous actions. Capital has high zones of manoeuvre in the form of commodified property rights, but is dependent upon milieux and nations (e.g. USA) in which those rights are enforced.

(8) Spatio–temporal flows create new spaces and times.

These are probed through innovations such as scenario-writing and focus groups to anticipate chronotopic tendencies and the future. The Fordist sequentiality of recurrent action patterns that we are used to is being transformed by incursive time.

Castells concludes that technology does not determine society. Society embodies technologies, yet societies do not determine technological innovation.

Strategic Co-ordination Information Technology Systems (SCITS)

Strategic co-ordination and control within the enterprise and between the enterprise and its value chain involves developing and connecting a distributed array of space–time cost information systems for managing the fluctuating dependencies between multiple, diverse and conflicting forms of knowledge. Much is known about the explicit knowledge platforms, their architecture and the modules that suit particular platforms. These are strategic co-ordination information technology systems (SCITS) and organizational innovations. Moreover, there is also considerable practical expertise in certain firms about the best elective affinity between SCITS and different ways of organizing process innovations.

SCITS provide the framework for the evolution from basically manual systems with simple mechanical and then electro-mechanical calculators that were then automated (Zuboff 1988). The MRP and MRP2 systems (see Chapter 8) provided the corporate centre with a vision of control while having the actuality of disconnected islands of data surrounded by impenetrable social structures through which social navigation was slow.

SCITS are clusters of innovations that embed and convey evolving information control and media technologies especially designed for firms in long chains and global networks: global commodity chains (GCCs). They are networks of power (Hughes 1983, 1987). The control and co-ordination revolution contains an array of cumulative capabilities that have evolved over almost two centuries commencing with the American System of Organization and Manufactures (Clark 1987):

- Standardized interchangeability of components
- System of design and abstractly conceptualized tooling
- High-performance products (e.g. rifles, field guns).

This very expensive capability was visioned in the early nineteenth century, delivered at certain sites in the mid-nineteenth century and became part of the social capital by the 1870s. Its capabilities were all too evident in the American Civil War. In the early twentieth century Henry Ford employed process engineers (1909–15) to:

- Develop and design a flow system for single, stable products
- Achieve space–time exactitude of pacing and synchronization
- High-product performance
- High-volume repetitive production
- Low cost for production and for revenue flows
- Mass distribution
- Vertical integration.

Fordism contained a number of problems in its interface with consumers who, although capable of customization, preferred General Motors' solution of the

- Multi-divisional decentralized organization collectively facing different market niches
- Sophisticated anticipation and choreographed future consumption
- Push, Just-in-Case push production systems protecting the firm from disruption (J.D. Thompson 1967).

Sloan-ism (1920–1960s) was a major organizational innovation in which a vast proportion of activity was transferred from humans into equipment for transforming, transferring and controlling activity. Abernathy (1978) profiled its core features in the American automobile industry from its founding into the early 1970s. The basic problem was that of high fixed costs, especially of the design of products and processes. European systems interfaced smaller market segments and their organizational innovations followed a different trajectory. Japanese firms in certain sectors developed major innovations. They also faced smaller market segments than did the American firms. Japanese automobile firms achieved:

- Multi-products
- Robust design with short lead times (e.g. Time-to-Market)
- Pull production systems
- Just-in-Time pull-based competition
- Low-cost inventories
- Incorporation of the supply chain into design
- Performance criteria of low-cost, exceptional quality, attractive aesthetics
- Tight synchronization (e.g. cellular manufacture for high volumes)
- Agile, ambidextrous organization
- Elective affinity with strategic co-ordination information technology systems.

In effect these cumulative developments create a cybernetic capacity in organizations (Best 1999). Japanese firms in consumer electronics were leaders in the development of mass customization. This included:

- Intensive design
- Extensive new product development
- The use of generic technologies to achieve high integration of different systems.

By the 1990s American firms, especially in SCITS were active in:

- Maintaining high performance, low cost and high quality
- Moving into robust, multi- and smart product design
- Flow system integration
- New system and process design
- System transitions.

Mass customization and SCITS achieve their competitive advantage by altering rather than entrenching innovations through organizational innovations in processes.

SCITS are infrastructures for decentralized technologies that connect widely dispersed geographical sites by combining common standards with situated, customized and flexible technologies (Star and Ruhdler 1996). The infrastructure is a web of usability and action for connecting activities through processes rather than structures. Contemporary forms of SCITS are a 'when' concept that is a foundational configuration that is embedded inside other structures, technologies and social arrangements. The systems possess a scope that includes individual events, giving them transparency in use and visibly supporting certain tasks. SCITS are learned through the membership of specific firms where SCITS are shaped by and also shape conventions in the community of practice (as in Lave, Wenger, Brown and

Cook). SCITS are most visible when they break down or are interrupted, possibly by a conflict between the global level and local circumstance. SCITS embody and potentially impose standards on other processes and systems. The architecture of SCITS is a potential source of inertia (Zuboff 1988). Equally, there is often a gap between the vision of SCITS by metropolitan suppliers and the users (see Chapter 9).

SCITS have three-fold spatio-temporal implications for material practices, for representations of space and the space–time representations (Sahay 1998).

(1) In material practices they facilitate the transfer of information through enhanced computer memory and storage, improving access to many others while raising the issue of access and data-ownership. SCITS are the medium for distanciation. Potentially, SCITS redefine existing norms of social interaction. Although some gain privileged access, there are issues of surveillance and exclusivity that impact on the shifting balance of power and changes in control.

(2) SCITS play a role in redefining measures of space–time by creating and manipulating maps and opening up new dimensions to perception by developing new skills. For example, the new systems of mapping introduce three-dimensional representations that allow access to 'forbidden' time–space domains and could impact on regional cultural spaces. There may be changing mental maps arising from new hierarchies. Unequal access unfolds.

(3) Representation gives new experiences of attraction/repulsion, presence/absence and inclusion/exclusion. Planners may construct new seductive utopian imaginary landscapes and new forms of presenting history. There is an issue of familiarity and unfamiliarity. Familiarity enables network mediated exchange that impacts humour and introduces hypertext notions. Unfamiliarity can result in a loss of a sense of property and a sense of repression.

The Internet is an organizational 'force' providing a material base for working. It enables flexibility in networks because they operate within and outside existing firms (bureaucracies). The network enterprise can assign goals to its networks. Operators are unlikely to be home-based because work will be multi-local in cars, offices and at home. So, the Internet may promote the de-centralization of urban areas. Cities might become vast mega-cities as in the example of the Southeastern quadrant of England centred on London. Some regions may have an elective affinity to SCITS that promotes Schumpeterian creative destruction over longish periods (e.g. San Jose, Silicon Valley and San Francisco). The Internet alters competitive advantage because innovation becomes a delicate blending of co-operation and competition as illustrated by the Linnux case, which is free to users, but they must shape the improvements. The competitive advantage is likely to be with those already wealthy nations whose social capital can accommodate the new organizational innovations in work and life-styles.

SCITS are likely to breach notions of social integration (e.g. collegiality) so central to the rhetoric of socialist ideals (Hodgson 1999). Those ideals fit poorly with the volatility of the Internet. Liquid modernity is a serious issue (see Bauman 2000). Also SCITS are likely to reduce the monopoly of the universities in knowledge diffusion because new suppliers can move in.

Space–Time: Commodified, Stretched and Colonized 3

The previous chapter introduced space–time compression and commodification as core processes in the new political economy (e.g. Figure 2.1). This chapter connects organizational innovation with the colonization of space–time by dominant agencies of various kinds. Organizational innovation involves the stretching of space–time from the local into the global. Yet, space–time colonization has been omitted in innovation theories despite considerable conceptual and analytical attention by the social sciences in the past two decades.

Organizational innovations occupy space and time. For example, the multi-divisional form emerged in the USA during the 1920s in a few firms, each of which was able to stretch its control over its activities beyond the local, beyond the USA and into the rest of the world (Chandler 1962, 1977). These firms commodified and stretched space–time from their domain of control in single cities (e.g. Boston, Cincinnati) to colonize the world. However, the necessary processual mechanisms and practices have not been sufficiently understood because analysis is narrowly confined to common-sense notions of space and time. Space and time in innovation studies are *only* viewed as objective, homogeneous units (e.g. distance in kilometres; time in seconds, days, years and time lines). These standard units can be added, subtracted, divided and multiplied, and therefore used in a vast array of calculations. The common-sense view is that these standardized units are not socially constructed and are therefore not cultural artifacts (see P.A. Clark 1975, 1985). This mistaken view has dominated both space–time geography (Glennie and Thrift 1996) and the structuration theory of Giddens (P.A. Clark 1985, 2000). Structuration theory influences the geography of diffusion and the analysis of corporate information systems by the Management Information Systems community. These expert systems are designed to store and transmit information across space–time as unchanging immutable mobiles. However, the position of place and temporality requires attention.

There are two main sections. First, there are four vignettes that explain and assess:

- Powerful heterogeneous actor networks
- Cathedrals as assemblages without maps (Turnbull 2000)
- The powerful gaze of the metropolitan professionals such as urban planners and industrial engineers who are most deeply involved in time–space colonization

- Giddens' space–time framework provides an important bridge between innovation theory, information systems and geography. Time–space stretching is a central mechanism in the development of cities, firms and nations. It is also a key element in the role of communication and information technologies within the globalized economy.

The stretching of space–time lies at the core of network innovations. Second, the conception of space and time as the only objective is wholly problematic. The colonization of space–time requires both culturally objectified frameworks and additional frameworks in which place–temporality possess quite different features. Moreover, given the hypothesis of mass customization and the requirement for organizational innovations that handle the future, then scenario-building and organizational narrative provide important potentials to be considered later.

Four vignettes

Powerful heterogeneous actor networks: PIEM framework

The process and politics of innovation require the creation of assemblages that produce obedience and handle resistance. Otherwise there is an unstable articulatory practice. The domination of a network depends upon extensive strategic alliances and truces grounded in recurrent action programmes that can cope with contingencies. Domination is the assemblage of networks of varied and distributed interests. Organizational networks should be conceptualized as the temporally stabilized outcome of earlier processes of enrolling other diverse actors. These earlier processes have become sedimented into the organization. Locating them requires an archaeology (running backwards) and genealogy (coming forward). A network 'maps' how agents translate phenomena into resources and then resources into networks. Bijker, Hughes and Pinch (1987; Bijker 1995) contend that networks are heterogeneous because in addition to people and their social relationships the networks include many non-human elements:

- Portuguese galleons (sixteenth century) as networks included the currents of the oceans, the winds, navigational knowledge, wooden planks, sails
- Cities contain many agencies that humans fear (e.g. rats, disease, automobiles) and accept (e.g. skyscrapers, advertising)
- The film *Jaws* included sharks, explosives, varieties of knowledge, resort owners, eaten holiday makers, policemen, heroes
- Formula-1 motor sport includes vehicles, tracks, drivers, designers, advertisers, rain, tyres, noise, many metals.

The puzzle to be unravelled is how these heterogeneous agents are enjoined into networks through a process Callon and Latour refer to as the *sociology of translation*.[1]

It is helpful to illustrate translation with the highly visible network known as McDonald's (Clark and Staunton 1989: 96–102). In the 1940s the original McDonald brothers sought to create a local franchise system in California but their franchisees did not maintain the categories. The franchisees substituted cheaper items and undermined the image by being less disciplined in their cleanliness. The brothers failed to dominate their heterogeneous actor network. Kroc, who became a key member of the firm, entered from outside and consciously sought to achieve domination over both the human and the non-human agents. Kroc was not familiar with either the theory of translation or its Machiavellian lineage, but implicitly deployed translation. Kroc sought out human agents 1,000 miles away in Des Moines whose discretion was easy to control and whose discretion could be motivated. Kroc also developed tight forms of surveillance (i.e. obligatory passage points). So, franchisees that broke the rules faced mobilization from the rest of the network. Within one decade the organizational innovation constituted through Kroc's agency possessed immense power and stretched across America. Kroc translated the non-humans, such as fats, potatoes, beef, equipment and buildings, through differentiating the network, to include a multiplicity of obligatory passage points (see shortly).

The sociology of translation enables the analyst and creator of organizational innovations to understand that successful, stretched networks are the outcome of processes in which principal agents (e.g. firms) author appeals to the political interests of other agents. The principal agent is active in socially translating other parties and phenomena into resources and into alliances that are structured networks of control.

This is the sociology of translation, with four moments that we shall refer to as *PIEM*. This will be illustrated from the role within Formula-1 motor racing of one of its principal agents, the entrepreneur, Bernie Eccleston. There are four key moments in the sociology of translation:

(1) *Problematization* [P] is the moment when agents attempt to enrol other agencies by proposing that the principal agents' definition of the solution-and-problems is indispensable to the other agencies. As in garbage can models, there is a tendency for solutions to seek problems (Cohen et al. 1972). Problematization involves constituting trials of credibility or 'obligatory passage points' as positions in the overall political process. There are tournaments between different solutions. All traffic must flow through obligatory passage points for network power to emerge (see Chapter 4). Otherwise the network

---

1   In the bibliography there are extensive references to the members of the research community. Useful introductions are provided by Bijker et al. (1987) and Bijker (1995).

lacks power and cannot stretch from the local into the regional and more distantly.

- Fifteen years ago Formula-1 (F-1) motor sport was a rather unattractive 'theatre of consumption' (see Chapter 5) with a spectacle that appealed to a small, enthusiastic local audience at the circuits in Europe. Wealthy enthusiasts and former drivers owned the teams. Even so, the sport was expensive. Eccleston began by introducing a series of new levels (see Chapter 4) and by enrolling agents at each of these new levels. In the co-evolution into a global media sport Eccleston translated others' interests in his network. The network principally included the TV companies which were looking for advertising revenue and which were directed at well-off watchers and owners of brands (e.g. Benetton) who were seeking to position those brands in the public consciousness. These three players (the sport, television and large corporations) were disconnected when Eccleston began by gaining a central power position in the sport itself. The organizational innovation consisted of articulating several levels across different time zones for media delivery to virtually all the globe, except for those watching Indy car racing in the USA. Over almost two decades Eccleston practised the politics of problematization and established quasi-legal obligatory passage points that the enrolled agencies accepted. Today the F-1 network is very influential, even though the spectacle of the sport seems somewhat ordinary and to be concentrated upon the start of the race rather than its completion. Only a minority of viewers watch the whole 90-minute spectacle.

(2) Interesting or *intéressement* [I] and enrolling other agents by locking in meanings and membership through the investment in some categories rather than others (see Thevenot 1984).

- In F-1 the locking in and the binding of parties involves forbidding any of the eleven racing teams to enter alternative championships such as Indy car racing in the USA. The binding in excludes the manufacturers and the advertisers, yet is sufficient to underpin the network at the level of the 18 or so global spectacles each season shown in global media television.

(3) Constructing alliances, truces and coalitions between the membership and the investment in categories to achieve *enrolment* [E]. Relationships have to be tightened and articulated into a consensually valid grammar (Weick 1969), through which all parties interpret their interests with the collective meanings and signification (Giddens 1984). The system of meanings (or habitus) acquires legitimacy and is the generative grammar (for the moment) of collective actions.

- F-1 has been immensely successful and all the agencies have benefited. Issues like safety and the survival of spectacular accidents gain legitimacy within the network and with the global TV consumers. There have been dramatic increases in the number of global viewers and therefore in the potential value to the firms investing in the teams through direct and indirect advertising.

(4) Ensuring that the network maintains its membership and meanings through *mobilization* [M].

- In F-1 the principal agent, Eccleston, makes efforts to fix the relationships among the enrolled parties (e.g. deals with TV firms) by creating political institutions.

It follows from the theory of translation [PIEM] that all successful organizational innovations, including the multi-divisional form, have probably been constituted in this way even though the players may have been unaware of their translation expertise. Moreover, given that these innovations are temporary stabilizations of many processes, it follows that their continuity has to be continually re-accomplished (Weick 2001) and that their diffusion (see Chapter 7) is a deeply political process of translating a multiplicity of heterogeneous actors. If so, then much of the analysis of innovation has omitted this crucial ingredient.

The theory of translation provides a useful insight into the ways in which leading consultancies create structured methodologies (see Chapter 8). These methodologies are likely to embody a double translation process within the consultancy and then from the consultancy into the client. Structured methodologies and their franchising to deliver systems such as Enterprise Resource Planning (ERP) imply the domination of suppliers inside their clients (with 'cookies') and among the users of the system. The theory of translation involves, therefore, examining the innovation process within client systems and existing approaches to implementation lack a theory of translation. They are, therefore, in principle, unlikely to work very effectively.

Cathedrals as assemblages without maps[2]

Why do some innovations travel so badly (see Jeremy 1981; Clark 1987)? This section emphasizes the *local origins* (i.e. placeness) of knowing and the significance of this fact for understanding and managing innovation. The metaphor for the local is the *assemblage* and its contrast is with the modernist metaphor of the map. The map is the modern exemplar of

---

2 This section draws upon David Turnbull (2000). The approaches of Law and Woolgar will be evident along with those of Callon and Latour.

scientific assemblage. The issue is: under which circumstances does the assemblage acquire maps and what hidden social and political mechanisms should be revealed?

The role of the local has been obscured by the illusion that there is a positionless vision of everything. That was overturned by the ontological turn towards heterogeneous actor networks whereby each site in a scientific community has a viewpoint. So, theories should be understood as boundary objects – a pattern that connects and separates different assemblages. Theories should be robust enough to enrol a community and malleable enough to embrace and inform local circumstances. Useful innovation theories need to possess the performative features of knowledge creation (see Nonaka and Takeuchi 1995).

So how was a cathedral like Chartres in northern France designed and built when the concept of a map was very cursory? How were innovations like flying buttresses imagined in the absence of a structural theory of mechanics? How were the huge variety of stones assembled and how were the different knowledges co-ordinated? Turnbull contends that cathedrals were built with talk, tradition and *templates* by masons, not architects. They were built in a performative approach at sites of experimental practice. The template or cursory sketch was the representational technology used to inform the cutting of stones. Templates, string, the compass and straight edges enabled radically innovative buildings. Thus there was an assemblage of messy knowledge that was made robust in a particular community of practitioners. Turnbull's analytic scope covers a diversity of knowledges and their assemblage. The assemblage process is missing from the highly influential contemporary theory of 'communities of practice' (e.g. Hutchins, Brown, Duguid, Cook, Suchman, Lave, Wenger and Weick).

It would be anachronistic to presume that there was a map for designing the cathedral. A plan or design is too strong a notion to express what happened. Instead, attention should be given to the co-ordination by masons of the various knowledges. There would be frequent daily consultations and the building process was discontinuous with its own periods of reflection. For example, the slow setting of the mortar and its possible shrinkage affected the daily rhythms, while there was also a building campaign.

Cathedrals possess some parallels to laboratories. Cathedrals were full-scale experiments – the innovations – whose outcome was closely scrutinized. Messy, heterogeneous knowledge was transformed into a more or less coherent outcome. Like scientific laboratories today, the medieval cathedrals absorbed large amounts of capital and concentrated resources. So, they became powerful loci of social transformation and constituted a system that was to some degree manipulable – as in experiments. The templates allowed for exactness and enabled a robust structure. The template organizes different work groups and accumulates stones into a building. Cathedrals were the focus and the engine for revolutions in work organization and in agriculture. They represented the resurgent integration of secular commerce and the sacred. Their construction also constituted specialized communities of tradesmen into a remarkable innovation in Europe. From that setting there was the gradual differentiation of the role of the architect in the seventeenth century.

Space is a contingent assemblage that can be represented by maps. The cartographic turn after the seventeenth century illustrates how maps combined with science and the state. Maps allow for enhanced connectivity and for the assemblage of information at centres of calculation. Their emergence as totems of scientific knowledge emerged from a configuration and complex socio-historical process in certain European states where the state, science and cartography became interwoven. They are the outcome of social action in particular knowledge spaces. Maps are mimetic representations of objective space. They are a prime scientific vehicle for framing and positioning scientific ideas. Maps are texts that can be deconstructed to reveal their concealed networks of power. How are maps equated with knowledge, including scientific knowledge? Mapping is fundamental to the processes of ordering.

One of the earliest attempts at map assemblage occurred in Spain and Portugal, but those attempts by the state failed to develop a coherent system from the previous form of naval charting. In each case the state faced the problem that the charts were rooted in maritime traditions that seemed to resist the imposition of a common grid of standardization and accumulation. The solution to grid-ing space and time was, of course, longitude and Mercator's projection (c. 1569). However the Iberian states were unable to construct that particular calculative framework and embracing assemblage. Longitude as the solution was developed in France and in England. Paris was the metropolitan centre from which a network of 400 triangles was constructed to map the nation by 1700. In 1783 Paris and Greenwich observatories mapped latitude and longitude as a grid. That level of activity is only possible when the state, science and cartography are integrated around exceptional knowledge-making capacities. Thus transnational space was created.

The European construction of maps and of longitude may be compared with systems of navigation used by the indigenous islanders of the Pacific. Lucy Suchman (1987) has celebrated their navigational knowledge in her account of plans and situated actions. Turnbull emphasizes the similarities between the political and social structures underpinning Pacific navigation without maps and the movement of medical and scientific knowledge from the laboratories into useful innovation theory.

The analytic problem is that of how the local is stretched into the extra-local through particular political systems of knowledge making. From the perspective of historically contingent constructivists there are no universal criteria of rationality and we must reject the notion of the individual as the rational actor. The new political economy contends that knowing is an interactive, political, contingent assemblage of space and knowledge that is sustained and created by social labour. Knowing is riddled with indeterminacy and all ordering is incomplete. Therefore there is equivalence between building of Chartres cathedral in the medieval age and the modern laboratory (Latour 1987; Law 1994). A cathedral built without maps expresses the claim that there are a variety of knowledges. In the assemblage perspective knowledge is less like believing statements and more like acquiring a skill. Therefore the role of models and structured methodologies becomes of particular interest (see Chapter 9).

The assemblage perspective challenges and is replacing Kuhn's (1970) view of knowledge as paradigms. Three decades ago Kuhn's theory of paradigms overturned the normative and representational model of science as an objective theory. Kuhn was the most successful at that time in drawing attention to the role of the community of practitioners in shaping the disciplinary matrix. Within the matrix are agreed examples of puzzles and solutions that are used to calibrate ongoing scientific activity. These are known as exemplars and they provide the ability to make equivalence between otherwise disparate problem situations and therefore to deploy the cognitive repertoire in a recognizable and communicable manner. Within the paradigms the models are artificial, finite constructs that although containing inconsistent tendencies are useful. Yet, Kuhn's paradigm expressed the theory of order as in singular forms of rather explicit knowledge. In contrast the assemblage complements the notion that scientifice activity is organized through partially open research programmes rather than paradigms (see Chapter 2).

The role of practice in science has been re-theorized since Kuhn drew attention to the role of the community of practitioners in shaping the disciplinary matrix. Kuhn's view of paradigm broke the Mertonian mould by transforming the normative and representational model of science as an objective theory. In Kuhn's paradigms, models are artificial, finite con-structs that, although containing inconsistent tendencies, are useful.

In order to contrast maps with assemblages and to show their con-nection it is necessary to confirm the local-ness of knowledge creation and the problematic of its transfer anywhere beyond that locality. Boisot (1998) and Nonaka and Takeuchi (1995) partly recognize the element of local-ness in their leading-edge theories of knowledge management, but they underplay the problems.

The laboratory is organized as an intentionally artificial rather than a natural setting. It is a controlled abstraction and extraction, enabling activities not possible in natural settings. Experiments are highly contrived. These performative features of science show that laboratories are special spaces constituted for the processes of assembling, standardizing, trans-mitting and utilizing knowledge. From the local space of the laboratory the distribution of knowledge is not guaranteed. Callon and Latour highlight the necessary politics and power relations necessary for that to happen. Therefore the distribution of knowledge has to be on the agenda.

The assemblage perspective on local spaces emphasizes the variety of knowledges, their situating and situatedness, and their contingency. More-over, all knowledge spaces are potential sites of conflict and resistance. Knowledge spaces contain a wide diversity of components (e.g. people, skills and equipment) that are connected through strategies and devices to constitute heterogeneous engineering. The local producers of knowledge pursue the strategy of seeking equivalence between their puzzle-solving regimes and those other knowledges that are otherwise potentially quite different. In order for knowledge to be accumulated there has to be a social investment in categories that is equivalent to any investment in hardware. So, social strategies in the local knowledge space attempt to locate the local knowledge in wider contexts through enrolling other agents, offering them seductive definitions of problem-solutions, creating trials of credibility and eventually – if successful – by creating powerful networks. These social processes require non-social devices whose feature is that they are immutable mobiles being transported through different knowledge communities. The devices and strategies of assemblage need to be mobile because the source of power in science is the ability to move local knowledge beyond its original site of production.

The robustness of findings is achieved in the face of antagonistic struggles through those immutable devices that are presentable, readable and can be combined with others at some distant point. The problem is resolved through creating several sorts of devices:

- Socially reliable groups of witnesses
- Journals and cognate publications that can carry diagrams and accounts
- Making technical apparatus reproducible and reliable.

However, science is about knowing one's way around the crucial site (e.g. laboratory, clinic, field site) and applying theory by recognizing that situations are similar. Scientific culture is deeply heterogeneous rather than homogeneous. Robust theorizing is an assemblage that is socially constructed in the face of contingency to embrace incompatible components. In that knowledge space otherwise disparate signals are *rendered equivalent*, are made more general and more cohesive.

Colonization: the powerful gaze of metropolitan professionals

Time–space colonization is one aspect of those activities of knowledge professionals who define the frontier of control for capitalism (Clark 2000: 139–140). They do not own the means of production, but aim to sell their expertise to the agents of the owners. The expertise they claim to possess has been acquired through state-approved processes of gaining credentials as members of professional associations and through higher education. Their expertise includes the capacity to conceptualize problems through their investment in certain languages and systems of categories. To varying degrees they develop a capacity to extend the influence of their cognitive styles and its discourse through colonizing everyday life. This will involve jurisdictional conflicts between professionals and conflicts with those whose everyday lives are being colonized. In some nations professionals are particularly influential (e.g. USA, Germany, France) and other nations rather less so (e.g. United Kingdom).

Typically, professionals structure the spaces and places of firms and cities. Foucault observes how the extensive zoning of time–space in relationships and practices emerged during the seventeenth- and eighteenth-century growth of the European nation state. The sphere of military activity was one in which new principles were extensively innovated, recorded and transmitted through the officer class. For example, West Point in the USA was the milieu of many organizational innovations in space–time control and was also the location from which the American civil engineers pioneered the control of waterways, canals and other natural features in the 'taming of the wilderness' (Clark 1987; Hoskin and Macve 1988). From the late nineteenth century onwards specialist professionals in the USA commodified their expertise about how to layout the flows of all kinds of work situation. Their expertise became one of the major templates for the best practice commodification of space–time. Space–time was treated as a commodified resource to be inventoried, controlled and observed. The retail sector was a major site for these activities and some American innovations were appropriated by British retailers (Clark and Starkey 1988).

Gregory (1994), the geographer, observes that professionals influence the relationship between abstract space and concrete space/place in cities. Lefebvre (1974/1991) contends that professionals enable capitalism to impose its tonalities through a process of colonization whereby abstract space is superimposed and hyperextended on top of place. Unravelling their influence on urban planning and life in the city was the major project of Lefebvre. In the critical geography perspective of Derek Gregory, these explain the gaze of power applied by the service class to restructure the *places*, *social rhythms* and *durations* of cities, work places and nations. One way of acknowledging their influence is through Foucault's notion of the eye of power (Gregory 1994: 401). This is illustrated in Figure 3.1.

Clark (1972) detailed a particular example of how the top management of John Players spent five years (1967–72) conceiving a future factory abstractly in time–space–cost before it was commissioned. Space–time was conceived as a cube without windows with only two open floors. Those space–times were located in the draftsmen's office on various charts. Time as a resource was related to the spatial cube to show the flow of work. The flow time would fall from 100 hours to fewer than ten, possibly five hours. This was a clear example of ablating whole sections of human activity and replacing it by new process technologies. The massively reduced workforce moved from day work into three shifts. On the very first day at the company and with the directors I was able to contrast my walk through ethnography of the existing factory and recognize certain key differences in social organization that would be inserted into the future factory. This is an example of time–space as a resource. The Directors of

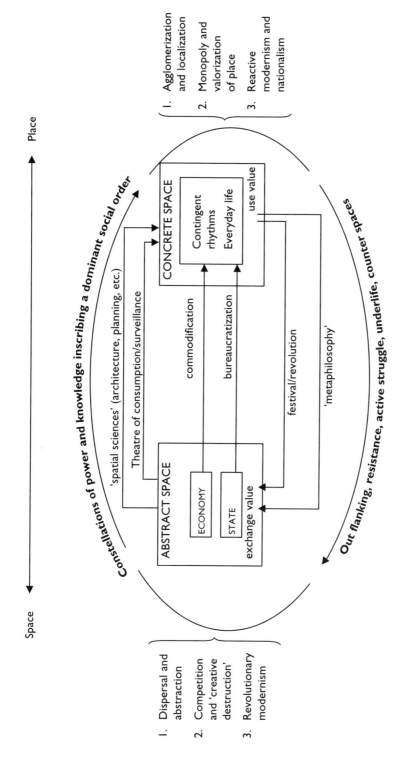

FIGURE 3.1 Space/place rhythms and the eye of power (modified from: Gregory 1994, Figure 30)

Players produced financial calculations to show how there were great financial savings of measured space and measured time in their design for the new factory. Morever the factory was designed to be 'the most technologically advanced in the world'. They used these arguments to persuade the Board of Imperial in Bristol that they should be allowed to build the new factory. In moments I was able to imagine how the new factory design would impact on the existing social organization and how it would inhibit the Board's desire to introduce autonomous work groups. The firm already possessed forms of autonomous work groups and management didn't know. Their proposed design relied on task groups. The consultancy revolved around two issues. What were the differences between the existing organizational economy of the factory and the proposed ideals? What alternatives were feasible in the lengthy time frame of several years before opening the new factory? This became known locally as the alternatives and differences approach (Clark 1972) and some of its features were appropriated by the architects and by the managing contractors as part of their repertoires for later projects.

Objective time–space and structuration

Giddens

Giddens adopts the widely accepted distinction introduced by the social historian, E.P. Thompson (1967), between pre-modern event time and modern standardized time and the claims that the former has been displaced and replaced by the latter (see P.A. Clark 1985; Glennie and Thrift 1996). This allows Giddens to incorporate the chronically recursive movement of individuals through time–space from the time–space geography of Hagerstrand (1978). Giddens also formulated an historical perspective based on the occurrence of events in the elapsing of calendrical time. History is the chronicling and explication of those events (Giddens 1984: 200–202). Giddens proposes to develop an historical and longitudinal perspective by drawing from the macro-historical project on world systems of Braudel. This operates on three different temporal registers:

- The glacial slow times of physical geography (e.g. climate) and its oscillations (e.g. ice ages). This is the *longue durée*
- The *moyenne durée* of 50-year phases in the growth and contraction of economic activities
- The short term events of days, weeks and seasons.

Braudel has been interpreted as a calendrical historian who superimposed additional time registers that were so long-term that their objective elements are obscured. Gurvitch (1964) famously contested Braudel's limited usage of the spectrum of social times (Clark 1975, 2000). Hall (1992) claims that Braudel uses objective time and space. My view is that the critique of Braudel by Gurvitch is significant. However, because Braudel sought to borrow from sociology, especially from Mauss (1904; Clark 1975), he provides a more variegated framework of temporalities than objective time. Braudel asserts the linking of everyday, weekly, monthly and other recursive patterns and social rhythms to the *longue durée* of social institutions. Giddens' usage of Braudel, however, does not suffi- ciently break from the common-sense view of space–time. So, the pivotal concept of 'chronic recursiveness' is insufficiently developed (see Clark 2000).

Giddens heavily emphasizes that modern society is characterized by declining presence-to-availability (e.g. Internet relationships) but not its curtailment. Giddens highlights the time–space stretching of societies into social chains that reach across places into spaces. There is a socially constructed bracketing of 'temporality–place'. Considerable attention is given to highlighting the role of *expert systems* that deploy global explicit and tacit knowledges to structure places that are distributed and contested. One key example of an expert system is the software systems for co- ordinating and controlling the resource flows of firms and their networks.

Societies possess distinct space–time cultural repertoires. Therefore there are time–space edges along which different societies encounter each other. For example, Clark (1997) contends that American and British space–time frameworks do differ. Consequently, the cross-cultural movement of British-American services, including leisure and bereavement firms, con- fronts both world-time and different time–space edges in the USA and in Britain.

Revising structuration

Giddens' approach to structuration requires revision:

- Thrift's (1983) revisions of Hagerstrand, Bourdieu and Bhaskar are ignored even though they have been influential in the geography of innovation. Giddens' time–space geography is transformed in the new geo-historical imagination
- The task-time of E.P. Thompson (1967) is insufficiently critiqued (Clark 1986)
- Time and space as analytic resources are under-developed (see Clark 1972)
- Neglect of Lefebvre's concern for the disappearance of lived time and of the spectrum of temporalities
- Too much focus upon 'extended present' and synchronicity
- Liminality requires more attention
- Under-developed treatment of time–space as resources within capital- ism and social market economies. Weak on political economy

- The continuity of practices presumes reflexivity, but reflexivity is only possible because of continuity of practices (Giddens 1984: 3).

These problems have been noted in a scattered, yet important array of critiques.

Since the mid-1980s the robust development of contextualist theories has provided an important and richer variety of time–space contributions. Orthodox approaches have been augmented and coupled with searching critiques of the previously established positions. However, the treatment of time, space and consuming remains under-analysed (see Chapter 5). Urry (1995) contends that places are increasingly reconstructed as centres of consumption and are increasingly consumed visually, yet that consumption becomes exhausted by use. In such places it becomes possible to consume one's identity amidst the contradictions and ambiguities of places. The spatial underpinnings of everyday life have been transformed yet these transformations are poorly accommodated in organizational theories. The consumption of space has become the economy of signs and these pervade both the cultural industries and increasingly all sectors of business and everyday life.

In the new political economy there is attention to the hidden role of time–space commodification in promoting a form of colonization through market power, leading to the 'compression' and the restructuring of local spaces and places (Gregory 1994). Obviously, changes in capitalism have led to revisions to the seminal Just-in-Case approach to organization theory articulated by J.D. Thompson (1967). Four lines of analysis are being developed:

- Geo-history and retroduction (Sayer 1992, 2000)
- Using the historical imagination to engage with the geographical imagination (e.g. Gregory, Harvey, Landes, Soja, Thrift)
- International competition between contexts and elective affinities (e.g. Porter, Castells, Sorge, Landes, Clark)
- Mobility of finance capital and its interface with the more inertial industrial capital (e.g. Harvey, Thrift).

There is a slow acceptance of the necessity for a more robust analysis of time and temporality and the requirement for a more depthful ontology (see the next section).

Space/place and time/temporalities

Towards depthful ontologies of process

It is now necessary to consider those theories of space/place and time/temporality that propose ontologies relevant to the process agenda. This

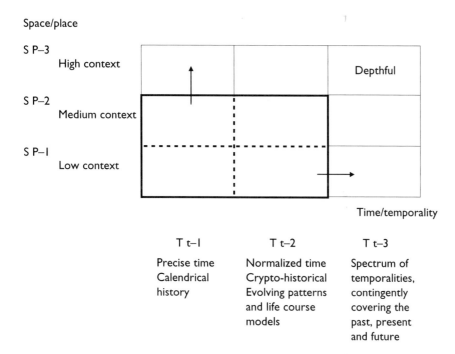

Space/place

S P–3
    High context

S P–2
    Medium context

S P–1
    Low context

Depthful

Time/temporality

T t–1

Precise time
Calendrical
history

T t–2

Normalized time
Crypto-historical
Evolving patterns
and life course
models

T t–3

Spectrum of
temporalities,
contingently
covering the
past, present
and future

FIGURE 3.2   Ontologies of space/place and time/temporality

section treats space and time as spectrums of social constructions that have
become objectified to varying degrees. Therefore common-sense ontologies
of space and time are simply part of the cultural repertoire rather than *all*
the cultural repertoire. More significantly, there are everyday taken-for-
granted usages of place and temporality that radically inform everyday
practices while differing from and silently co-existing with common-sense
ontologies.

   To inform the examination of organizational innovations the dimensions
of space–place and time–temporality are trichotomized into three main
research programmes. This creates a three-by-three grid (Figure 3.2) that
stylizes the main issues which should be used pragmatically to deepen
analysis. The aim is to contrast the box (left-hand quadrant) containing
'real' and normalized space–time with the notion of depthful ontologies
shown in the top right. We start with time and then temporality before
examining space and place.

Progressive Time in Action: 'Historical, longitudinal . . . real-time'

This section problematizes the position of the 'historical, longitudinal,
long-term and real-time' in the positivist notion of progressive time. In

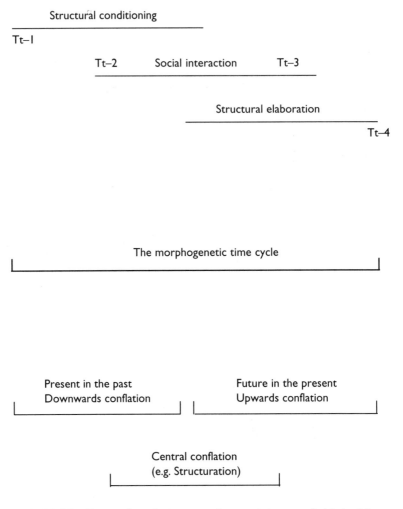

FIGURE 3.3   Temporal conflation: central, up and downwards (derived from: Archer 1995)

innovation studies there are many references to dynamic processual approaches, especially the variants of evolutionary dynamics, and the 'in time' school of analysts. Linearity and homogeneous time is implicated in all these. Positivism compresses processes because time is treated as a homogeneous variable (e.g. Mohr 1982; Miles and Huberman 1994). Consequently, the position of temporality and the plurality of heterogeneous event-time reckoning is not thoroughly grasped. Therefore, the full significance of the anti-modern contribution containing the post-modern, the realist turn and structuration perspectives is sidelined. They reject the imposition of grand narratives as in Marx's analysis of the dynamics of capital and the civil society. How long is the long term and how are processes periodized? What is meant by the long term? Is the long term centuries (e.g. Braudel), generations or three years as in concurrent 'real time' studies? The problem is how to re-theorize linearity.

Much of theory and research on process examines events within an extended present. There is a strong tendency to ignore the post-modern challenge and its claim that history is theoretical rather than simply factual. So, business histories may leave ideological assumptions unexplored. Frequently the time frame is only in calendar time. Even studies that claim to be longitudinal cover only two to three years – the time of a doctoral thesis. The references to 'real time' are less impressive than is suggested because the temporal sequencing of actions tends to be obscured within an extended present that addresses neither the possible influence of the pre-existing nor the dislocations that might unfold in the future. There are four basic problems:

(1) The quasi-historical case studies often leap from the founding period to the present. It is rare to find an analytically sound periodization of episodes. Case histories often exaggerate the role of high-status persons, especially in the 'turnaround type' of analysis. Typically the narrative is detached from the underlying mechanisms and the influence of the context. There is a surface narrative of modernist achievement.

(2) Life cycle, multi-stage theories are widespread and redolent of the development theories from the modernist period. Consequently, there is too much attention to the notion that future events can be ordered and controlled. Abernathy's (1978) critique of sector life cycle models is a starting point for inserting a processual understanding of temporality (Clark and Starkey 1988).

(3) Foucauldian discourse analysis is used to good effect by Townley (1994) in her critical political economy of human resource management. However, discourse should locate the struggles sedimented by discourse into truces. Hoskin and Macve's (1988) analysis of West Point Academy insufficiently unpacks relational configuration of groups and interests involved in the political economy of American conflict with Britain.

(4) The future cannot be extrapolated from the past because the previous cycles of recursive structuring are dynamically contingent.

The new political economy advocates the analytically structured narrative *sans grandeur* with case-like cameos in which the temporality of events and the placeness of spatiality are implicated.

The position of linearity is challenged by the post-modern critiques. So, the notion of history as a common-sense empirical account of what actually happened has been widely challenged and not only by the post-modern turn. Those following the realist turn promote a view of temporality that aims to make the enduring elements of structures and causal mechanisms contingent.

ne and Temporalities

e time dimension of organizing innovation was initially dealt with
ough very abstracted and almost friction-free time-lines from an
specified past into a utopian future. In contrast, the current research
programme emphasizes that there is more inertia and transformation.
Moreover, transformation is unevenly distributed so that even nearby
regions are quite differently affected. Further, there is growing attention –
and not just in complexity theory – to the dynamics of morphogenesis
(Clark 2000). These analytic developments in eclectic theory suggest that
the simplistic treatment of 'history' is unsatisfactory. For example, in the
Academy of Management it is taken for granted that time is now the hot
spot for developments in theory, research and practice. However, the
dominance of progressive time undermines and vitiates the development of
the process agenda and requires revision. We should therefore distinguish
three ontologies of time (Tt–1 and Tt–2) and temporality (Tt–3):

Tt–1   Precise time in standardized units from the clock and calendar used
       to measure resources, to express flows and construct time-lines.
       Precise time was a focus of revived interest after innovations such as
       Just-in-Time, Time-to-Market and Time-based Competition.
Tt–2   Normalized time, timeless process and frictionless ontologies. This
       has been the major focus of attention for three decades.
Tt–3   Spectrum of temporalities dynamically and contingently recursive
       in the past, present and future.

These are shown on the horizontal axis of Figure 3.2. More than 90 per
cent of citations use precise and normalized time.

First, *precise time (Tt–1)* is one of the taken-for-granted forms of time
with no attention to temporality. This is the salient, but not the only time
frame of contemporary corporate planning. Firms are treated as temporal
inventories and time is commodified in apparently precise calculations.
Time becomes a corporate resource because, like money, time is a core
framework in political systems. Precise time in the form of the calendar has
also been heavily invoked to provide a gap between the past and present
and between the present and future. These forms of time lie in the heartland of SCITS (e.g. Timetables, MRP2, ERPS and logistics). Time-lines
have been influential in the geography of innovation, in corporate histories,
in long-wave theories (e.g. Kondratiev, Schumpeter, Mandel and Freeman)
and in Braudel's usage of the 50-year economic pulsation. Current attention to time is still heavily grounded in precise units.

Second, the notion of *normalized time and time-boundedness (Tt–2)*
currently occupies the core of existing theories. The aim is to avoid the
imprisonment and specificity of precise time while still displaying the
analytic mastery of past, present and future events typical of progressive
time. Normalized time is an abstraction from situations. This is also the
time of SCITS. Normalized time can be in cycles (e.g. firm life cycles, sector

cycles) and/or in linear, non-reversible phase models with a succession of stages, as found in many innovations models (Wilson 1992). The linear sequence contains definite episodes in the form of the succession of synchronic moments (Daft 1997). Equally so in the path dependency models (Scarbrough 1998). It has been argued that these are a synchronic illusion. Their influence is found in the realist theories of morphogenesis and in the theories of the learning organization (e.g. Senge). Life cycle and multi-stage models typify normalized time and time-boundedness. They consist of a series of cross-sectional forms that are motionless. They therefore do not deal with recursiveness either in the ongoing events of stasis or in transformation. Consequently, repetition and difference are glossed over. Van de Ven and Poole (1995) rightly retained this form of time in their critique of change theories, but a depthful ontology requires a more central position for events and temporality. The current state of Tt–2 is still exemplified in the approaches of Weick and of Barley. Weick (1969, 2001) is a key author in offering an image of flows and of evolutionary transitions, yet Weick observed to a meeting of the Academy of Management (Toronto 2000) that he had not resolved the issue of an exquisite sense of time (see Weick 1969: 64). Barley's time relies on structuration yet escapes from some of the problems of objectified, normalized time (Dubinskas 1988). Barley's frame of reference is central to Orlikowski's seminal account of technology and structure.[3] It seems that 'Barley time' is the protective belt in information studies and organization theory. Barley time extends Giddens time, but does not yet sufficiently interface with event temporalities. Normalized time is prevalent and dominant. However, its problematic features are increasingly the focus of attention.

Third, *event filled temporality is very undeveloped (Tt–3).*[4] Critique of time in management as an object, of time compression and of real time. There are a variety of time scapes that constitute a multi-layered and coexisting spectrum linking production and consumption. Recently some historians have questioned the degree of certainty presumed by historians (Schama 1991) and proposed many possible futures (Hawthorne 1991) and introduced different, earlier outcomes, as in Ferguson's (1997) virtual history. Could the past have emerged differently? The issue of alternative pasts and uncertain futures directly challenges the claims of the realist turn to be able to estimate the degrees of freedom and hence the zones of manoeuvre available in specific unfolding situations. The framework of Gurvitch permits careful attention to the various possibilities. The new

---

3 I am indebted to Wanda Orlikowski and JoAnne Yates for their comments on an earlier presentation.

4 My position is set out in series of papers on the spectrum of times in modern societies (see Clark 1976, 1978, 1985, 1990, 1997, 2000). This position is close to that of Glennie and Thrift (1996) and we all differ with Giddens (see Orlikowski, Barley). There is an extensive, specialized and very informative literature in *Time and Society* published by Sage.

political economy requires critical narratives that reveal the possibilities of struggle. Zeldin (1994) contends that history can provide a detachment to analysis of the present so that a wider range of choices about the future can be imagined.

With respect to the future, how are possibilities, dis-locations, scenarios and virtual history to be understood? One issue for temporality is the scope of the past, present and future required in specific pieces of analysis. Temporality requires attention to forms of periodization embedded in the events and therefore indicative of the processes being analysed (P.A. Clark, 1985). Roth (1963) and Dubinskas (1988) illustrate temporality from the medical and scientific fields, revealing the problematic use of temporality in some studies of innovation in organization and technology. The requirement is both to establish a periodization of the past and to detect turning points, and also to apply some analytic notion of causality to selected and theorized events while being aware of the ways in which enduring structures can be de-institutionalized.

The potentials for periodizing and scoping temporality involve the dimensions found in the spectrum of social times proposed by Gurvitch (1964). I have computed his eight-fold taxonomy into three dimensions as Table 3.1 (P.A. Clark 1975, 1985: 44). The three dimensions are:

- The relation between the past, present and future
- The degree of continuity, contingency and surprise
- The durational experience, including the sense of pace.

Using these dimensions and types as an analytic tool should scope the temporality of institutions and structuration.

The temporality proposed by Gurvitch provides a strong challenge to the claim by evolutionary historians to have satisfactorily applied an institutional perspective to narratives through the models of cumulative path dependency, lock-in and trajectory. Their applications seem more like imprisoned time without attention to irony and emergence, and the critical search for alternatives is suppressed (see Ferguson 1997). Contingency and the unexpected are slotted into quasi-equilibrium thinking. Cumulative path dependency does not unpack recursiveness, ignores failures and produces a single variety. The possibilities for discontinuity, dislocation and revolution require attention.

Space and place

The issue is the relative role of place and space. Until recently innovation theories ignored the spatial dimensions by using frameworks that bracket and normalize space out of consideration. The post-modern contribution has significantly re-opened attention to the extensiveness of innovations and space. Post-modern perspectives reveal the role of multinational firms in commodifying, controlling and choreographing space for purposes of

TABLE 3.1 Gurvitch's spectrum of social time (from: Clark 1985)

Type	Relation of past, present and future	Continuity, contingency and surprise	Duration, incorporate pace
Enduring	Past is projected in the present and future. Remote past is dominant.	Most continuous.	Slowed down, long duration. Present can be quantitatively expressed.
Deceptive	Rupture between past and present.	Discontinuity. Surprise time.	Seems like enduring, but sudden crisis. Paradox. Simultaneously slow and agitated.
Erratic	Present appears to prevail over the past and future.	Uncertainty and accentuated contingency. Discontinuity becomes prominent.	Irregular pulsation between appearance and disappearance of rhythms.
Cyclical	Each is mutually projected into the other.	Continuity accentuated. Contingency weakened.	'A dance on one spot'. Qualitative element strong.
Retarded	Future actualized in present.	No equilibrium between continuity and discontinuity. Contingent elements are reinforced.	Delayed, waiting for unfolding.
Alternating	Past and future compete in the present.	Discontinuity stronger than continuity. Contingency not exaggerated.	Alternating between delay and advance. Qualitative not accentuated.
Pushing forward	Future becomes present.	Discontinuity. Contingency.	Time in advance of itself (e.g. communions in revolt). Qualitative.
Explosive	Present and past dissolved. Creation of the immediately transcended future.	Discontinuity high. Contingency high.	Fast movements. Effervescent. Qualitative high.

corporate interests and shareholder profits (Jameson 1991; Baudrillard 1975). The many analyses of the American shopping mall and of real estate in Orange County, California (e.g. Sorkin 1992) contend that leading corporations deploy the agency of marketing and branding to transform space into 'theatres of consumption'. The making of the American consumer involves Hollywood-ization, Disney-ization and the imagineering of everyday life through media-scapes. The 'scapes' are inserted into all areas: eating, drinking, leisure, shopping and even theatres. So, there is increasing interest in how to analyse places. For example, places contain distinct social templates and isolating mechanisms because location impacts on network creation.

The observation by Baudrillard that these 'new realities' of space have no real origin misses the point and hardly empowers the American consumer. This is because in the American context the notion of history and of an original template to be re-produced had already been overturned by the first celebration of the American Revolution in 1876. Place was soon transformed into a narrative of new spaces and of technologies in dams, skyscrapers and electrification (Nye 1997). By the mid-nineteenth century Americans used museums to show expositions of the future (Butsch 2000). Massumi's comment that Baudrillard misunderstood the supermarket provides a wake-up call.

The three dominant distinctions about the usage of space and place are shown on the vertical axis of Figure 3.2. First, the bottom level refers to the *common-sense ontologies of standardized spaces (SP–1) that are place-free*. This level is known as low context theories (Child 2000) and they dominated innovation theory until the late-1980s (e.g. Rogers 1962; Hagerstrand 1978). Hickson et al. (1974) contended that societal characteristics were 'washed out' in systematic, measurement research. This level excludes attention to the comparative management level (SP–2) and to the international political economy of competition (SP–3). In SP–1 space is conceptualized in standardized units (e.g. linear kilometres and cubic metres) which are abstract, codified and totally portable. The standardized approaches of representational knowledge can be used to construct maps and diagrams of places. There is a whole division of specialist occupations contributing to this discourse.

Second, *Comparative International Management (SP–2)*. World space is disaggregated into vast national spaces, each of which is treated as homogenized with respect to specificities of culture as in comparative international management. This massively influential level was stimulated by Hofstede (1980) who demonstrated what were measurable cultural differences between the scores of approximately 40 nations on feminism, collectivism, power–distance, and time frames. Hofstede's studies problematized the nation as a unit of analysis and its explanatory methodology readily and rapidly enrolled North American researchers (see Chapter 10). There has been a massive attempt to develop theories of medium context. They tend to highlight the distinctive features of particular nations, yet not to relate those differences to the competitive advantage of contexts. However, the very influential institutional theory of innovation developed by DiMaggio, Powell and Scott largely ignored the exceptional place and nationhood of the USA. Consequently, meso-level theories such as co-evolution and adaptation/selection also largely 'wash out' place and geography.

A significant part of this level is implicitly prescriptive and provides ways in which management can recognize the everyday differences between their own cultural patterns and those of other nations. The potentials of this level for understanding the dynamics and evolution of process innovation were prised open in studies that sought to situate national sectors in their competitive context of the world (Pavitt 1980; Porter 1998). This established the national systems of innovation approach (see Chapter 6). Much of this was stimulated by American concern about Japan and by European concerns about the USA (Clark 1987, 2000; Whittington and Mayer 2000).

I refer to this level as medium context (cf. Child 2000). There are several problems and these stimulated the development of the next level.

Third, the *New Political Economy: Competition between Contexts (SP–3)*. The development of high context new political economy theories centres on examining the relative influence in the world economy of three advanced nations: the USA, Japan and Germany. High spatial context perspectives embrace the medium level and situate the firm, sector, region and cluster within the international political economy (Porter 1990, 1998; Clark 2000). This rhetoric proclaims 'competition between contexts' as a primary analytical cut to any and almost every analysis of firms. This level is already fully occupied by geographers (e.g. Dunning 1993; Dicken 1998). There is a growing interest in organizational geography. Exponents of this level are drawn into a debate with those who would confine their analyses to the first levels. Examples of analysis at this level, which also involve time and temporality, tend to be stronger on time than the temporality, for example, in Braudel, Wallerstein and the recent development of the Political Economy of World Systems (PEWS).

An essential ingredient is to situate the firm in the context of social capital and the national market and to see that base as possessing a potential in the international political economy. This means colonizing the interface to international management.

So far, the high context approaches to space/place move towards a more depthful ontology yet often omit attention to temporal dynamics.

Depthful ontologies

The process agenda requires space/time and place/temporality to be situated in their relational configurations. Old questions need to be reformulated.[5] Attention to experienced distance arises in new questions of space/place. Do information technologies (inter- and intranet) both stretch time–space and also reduce distances? Do certain organizational processes become immobile and how does experienced distance define competitors (e.g. boundaries of nation)? How do successful multinationals manage distances and how does distance affect managerial processes (e.g. supplier chains)? If space is a cognitive representation, then does foreign competition mean that overseas rivals are closer? How do spatial features affect knowledge distribution in the global commodity chains that supply metropolitan supermarkets?

---

5   These topics were suggested and extensively explored in the special sub-group on space at the *European Group on Organization Studies* (EGOS) at the 2001 Symposium, Lyon, France. Their programme of 20 papers is on the EGOS website.

Multi-level Configurations: Their Stasis and Transformation 4

T his chapter examines multi-level systems and their implications for useful innovation theory. One problem is how to understand and manage both the re-production of existing systems and their alteration. Various frameworks, including complexity theory and structuration theory, seem to provide analytic solutions. However, while each of these does provide a more robust analytic approach than a framework based on variables, there is much revision and refinement to be undertaken (Mohr 1982). The move away from grand, law-like narratives towards the use of multiple frameworks requires an eclectic theory that can construct analytically structured narratives.

This chapter examines multi-level configurations and introduces the eclectic theory required for understanding and managing their processes. These include the analytic reconstruction of specific instances of *dynamic organized structures* (Gibbons et al. 1994) in terms of their past, present and future orientations. The challenge to useful innovation theory is how to cope with the actuality of dynamic capabilities in firms, regions, sectors, cities, nations and similar places. A first step is to develop a cluster of frameworks – the eclectic theory – that are processual and can cope with specificities of firms, regions, nations and the global economy.

Useful innovation theory has to address the issue of whether the existing situation can be transformed towards some ideal or theoretical prescription. Universal theories have obfuscated the issue of what are the zones of manoeuvre available in particular contexts for innovation. For example, American funeral firms entering the British funeral sector will be constrained by the difference between the American cultural repertoire based on open caskets and burial when faced by the British cultural repertoire of closed caskets and cremation. These are issues of 'what is agency?' and 'how is agency constrained by pre-existing systems?' Innovation theories have shifted from presuming a high degree of agency (e.g. creativity theories) to an emphasis upon the historical constraints of forward feedback and lock-ins that create finite paths and trajectories of innovation. Path dependency has been a useful corrective but is also limited. Hence the interest in and attention to the *interplay* of agency and pre-existing structural repertoires.

The problem is the relationship between agency and structure (see Figure 1.1). The analytic problem has been addressed by Marxists, postmodernists, post-structuralists, realists and by structuration approaches.

Giddens' (1984, 1990) theory of structuration is salient and offers a highly respected contribution to issues of time–space. Post-modernists successfully led the critique of positivism and the development of critical political economy. Their position was challenged after the rise of positive political economy (e.g. Reagan, Thatcher). Moreover, complexity theory has a strong constituency.

Earlier debates around multi-level theories are being reformulated into a new political economy possessing both a positive view of capitalism and a critical political economy of capitalism (Figure 1.1 on page 3). Each wing (positive and critical) requires an understanding of the multiple circuits of power and obligatory passage points (OPP) through which innovation is undertaken (Clegg 1989). The notion of multi-level circuits of power needs developing. Whereas theories previously presumed that innovation should be free from struggle, contemporary theorizing highlights the politics of innovation along many dimensions. Contemporary theorizing positions the users and opponents of specific innovations (e.g. to genetically modified foods). We need to examine the sites of action, such as in hospitals, schools and in bereavement, as well as more orthodox areas of work.

Ontologies for systems and people

Discourse and practice shape what it is possible to think and say. Discourse practice is a set of rules and procedures governing thinking, performing and writing in a given field (e.g. innovation theory). In every institution, discourse is the medium of ordering and governing because discourse is inseparable from power. Theories must compromise with the power consensus because individuals working within a particular discursive practice are pressured to obey the unspoken 'archives' and repertoires of rules and constraints. Yet we rarely know that archive until it is disrupted. This point explains the ambitions of the post-moderns to make the discursive practices of contemporary capitalism more transparent. In order to develop the process agenda, the ontological differences between the modern and the new political economy have to be articulated.

The issue of ontologies refers to the background beliefs about reality and to whether you hold that the acknowledging background beliefs about reality should take precedence over claims about epistemology. The ontological question is 'what constitutes reality?' All analysis of reality is mediated by the theoretical framing we insert between reality and ourselves. Therefore the social ontology that is chosen plays a powerful role in shaping and, in effect, regulating the explanatory methodology because ontologies govern the choice of concepts that are admissible in description and hence in explanations of reality. How do different ontologies infuse explanation? Archer makes three points. Ontologies:

- Are *theory laden*
- Occupy a regulatory role in governing *which concepts are admissible* in description and in explanation
- Act as a *gatekeeper for methodology.*

There are considerable ontological differences between the modern ontologies and those of new political economy. The current situation is one in which there is an ongoing three-way struggle between the ontologies of the late modern, the anti-modern and new political economy. It is assumed that each ontology retains exponents and advocates for some time after the emergence of rival research programmes and new theoretical positions.

First, in the modern era (1870s–1950s) the system analogies were closed off from the context, as in the case of the clock. The core analogies were of designed, artificial systems with an architecture of movement and components that were tightly connected in a purposive arrangement. Such systems require human designers and human agents to maintain the system performance. They are illustrated from the contemporary world of leisure by the swimming pool robot cleaner (Clark 2000). Closed, linear system models are best represented in the Marxian model of an economic base and political superstructure animated by the utopian evolution of capitalism. Boulding (1956) demonstrated that closed system analogies gained a strong grip throughout the modern period, even when alternatives were imagined. In the modern era economics provided the rational, purposive framework to theories of organizing, particularly in the various schools within Scientific Management. Neo-rational thinking also shaped the theorizing of bureaucracy as a political and organizational form. Rational economic theorizing possessed a strong influence, especially in models of human motivation. There was a cognitive dynamic (e.g. cognitive dissonance). The human relations movement provided an emotional, non-rational model. However, the theory of innovation was subordinated to a theory of economic equilibrium and Schumpeterian notions were largely ignored. There was little attention to organizational innovations. Indeed, there were latent stereotypes of Luddism and of resistance to change. Technological innovation was ascribed to tools, equipment and to workflows.

The late modern era commenced with the prising open of the closed system model. The imagining of open systems became a major project in the political and social sciences. This led to a struggle over the application of open system models to organization and to innovation. In open system models order can be unplanned and emergent. Theories of innovation were taken very seriously, but were tightly concentrated upon codifying the dimensions of technology hardware and constructing abstract taxonomies. Late modernity witnessed the flowering of expert approaches to organizing innovation. Although its position was continuously challenged during and after the 1970s, many of its research programmes were carried forward in what might be generously labelled, *thoughtful and thinking positivism.* The limits of late modernity had been identified, especially in the separation between the individual and the system. Also, there were considerable problems of coping with uncertainty and complexity.

The ontological struggle became one of *process versus structure*. The ontology of late modernity declared process to be so varied that its movements could not be analysed (e.g. Pugh and Hickson 1976; cf. Clark 1987). So, structure was presented as a stable, enduring configuration of dimensions: centralization, formalization, specialization, standardization and occupational profiles (Pugh and Hickson 1976; Mintzberg 1978). For a decade the Aston Programme was the leading edge. Although Hall (1972) sought to connect structure and process, that research programme failed to enrol followers, especially in North America. In Europe, Crozier sought to develop a universal processual perspective capable of detecting those features of innovation that were distinctly French. However, Perrow transformed Crozier's analysis of process and innovation solely into a universal model (e.g. Perrow 1967; Clark 1979).

By the late 1970s the *anti-modern theories* were beginning to flourish. The anti-modern is a collection of different approaches that share a critique and rejection of the modern. Clegg, Hardy and Nord (1996) demonstrate that the critique of the modern is still underway. Within the anti-modern there are different lines of critique: structuration and symbolic interaction theory, the realist turn (Clark 2000) the post-modern and post-structuralist (e.g. Clegg et al. 1996). The anti-modern ontologies introduced a quite different reality and theory of knowledge. By the 1980s the post-modern movement was pastiching and cutting through the neat divisions of late modernity. Hence the three key movements were providing new research programmes. Also, complexity theory revived the role of history while structuration theory both highlighted time and also became marooned in the extended present. In this era, organizational innovations moved from the periphery to equality with technological artifacts. Theories sought to go beyond organization choice (Trist et al. 1963).

The struggle between modern and anti-modern is superbly illustrated and explained in the *Handbook of Organisation Studies* by Clegg, Hardy and Nord (1996). This is a revealing compilation by the good and the great. The editors montage the current state of the (Euro-centred) anti-modern community and its debate with the modernist legacy. The montage enables reflection and highlights a series of problems. However, there are important problems:

- The simple debate between modern and anti-modern dominates. The modern is rather stereotyped. The unwitting moderns are let off and there is a limited account of competing communities of practice. Contrary positions are ridiculed
- The orthodox bridge between structure and process is relatively unexamined
- There is a lack of attention to mechanisms of social change or stability

69

- It does not fully address context or temporality. Its sense of geo-historical explanations is undeveloped. The chapter on time and the scattered references to space and place are revealing
- The area of consumption is not developed
- It fails to situate international political economy
- American exceptionalism in the international political economy is ignored
- Slight attention is given to the role of knowledge and of innovation.

These limitations, however, highlight the agenda (see Chapter 1) for going beyond an anti-modern critique and they illustrate the difficulties to be encountered in developing the new political economy of organizational innovations.

The new political economy commenced by problematizing the solutions of the anti-modern era. Alexander (1995) proposed to construct dynamic meta-theories that spanned multiple levels. Moreover, in the 1990s, organization theory proclaimed the meso-level and revised theories addressed the co-evolution of firms and their contexts (see Chapter 6). Also, the modernist certainties of planned change were addressed by formulating dilemmas such as innovation/efficiency (Abernathy 1978) and adaptation/selection (Aldrich 1999). Additionally, the hypothesis of mass customization built on the post-modern 'history' of the USA as a consumer society (Slater and Tonkiss 2001). *Relational configurational* theories replaced a variable approach to the theorizing of process (see Mohr 1982). Gradually, the previous satisfaction with structuration and complexity theories is being re-appraised. These shifts and their inherent difficulties are neatly compressed and expressed in the theory of morphogenesis proposed by Archer (1995).

Within the study of innovation processes the development of a multi-level relational configuration approach has been influenced by *complexity theory*. Complex systems are an intricate mixture of regularity and irregularity and of order and disorder. They are self-organizing dissipative structures with positive feedback loops possessing high levels of non-linearity between causes and effects, between inputs and outputs. They are in a sensitive dependence on the initial starting conditions and consequently small events can take on a major importance. The analytic problem is to understand the journey and trajectory of a system (e.g. firm-and-context). Complexity theory draws attention to the importance of history through the concepts of initial conditions, lock-in and path dependency. It is necessary to study trajectories and to acknowledge the role of chance events. However, understanding of the context is impacted by imperfect information arising from the mis-enactment of the context. Management can influence some trajectories, but only rarely control them in a detailed way. The behaviour of complex systems is held in a relational configuration by system attractors. The attractors bind the system into regular patterns with irregular performances. However, strange attractors have multiple points of attraction and these can unravel a complex system.

Analysts have to search for discoverable mechanisms that operate in a preordained way with bounded stability.

Using complexity theory should enable the better mapping of systems and their contexts. It may be possible to anticipate the alteration of attractors and hence the unbinding of a system. Complexity theory highlights the emergent possibilities. In complexity theory understanding complex systems means:

- Giving less attention to outcomes and more attention to the event cycles and processes in which outcomes are implicated in order to explain the relational configurations
- Periodizing of events to identify the key episodes, to examine durations and pace, to locate the causal loops
- Acknowledging the particularity of fundamental principles and generalizations
- Examining the conditions, tendencies and mechanisms that sustain particular generalizations
- Considering whether these mechanisms, conditions and tendencies will continue into the future
- Not assuming that adaptation by existing systems can be taken for granted as in much of the innovation and change literature
- Focusing upon the dynamics of external selection versus internal adaptation for the firm.

These features shift the locus and focus of analysis to agent-based models that give managers the capacity to model situations. That means that following popular notions such as 'best practice' is like watching a railway train heading for a crash. There is a plurality of best practices. Therefore we need to identify the conditions under which different forms of best practice have particular effects. It follows that we should expect innovations to be hybridized. Organizers need models that acknowledge idiosyncratic and emergent conditions. Because organizers use and consume metaphors, they should be shown how they can detect and map their metaphors. To assist organizers, they require subtle languages to examine time–space dynamics. Also, they need to develop design philosophies to help them understand dynamic interaction and to ask the best questions.

The interplay of structure and agency

Theory of structural activation

Organizations are theorized as possessing repertoires of recurrent action patterns (Clark 2000). The contents of the repertoire embrace dynamic sequences in various degrees of dormancy. Potentially, the sequences of recurrent action patterns are capable of activation, recomposition and assembly into new processes. This interpretation is consistent with the realist turn and with revisions to structuration theory to enhance temporality. The interpretation also confronts every theory that assumes that the firm can be ambidextrous and agile. The existing repertoire cannot be transformed quickly through frictionless change. Those assumptions ignore the finite capabilities of the repertoire of recurrent action programmes possessed by the organization. Finite capabilities mean that there are limited zones of manoeuvre. Firms have finite and emergent capacities that constitute zones of manoeuvre that are more limited than voluntaristic approaches have suggested. The focus upon the activation of structural repertoires theorizes that the repertoires pre-exist their activation and that the activation possesses emergent potentials.

Attention to the processual dynamics in organization studies has developed considerably, although the handling of temporality requires development (see Chapter 3). Clark (2000) contends that morphogenesis should be anchored in the concepts of repertoires, recursiveness, recurrent actions patterns, strategic time reckoning and repertoire activation. Therefore, as illustrated in Figure 4.1, it is necessary to distinguish between whether the repertoire remains the same through many completed cycles of its scope and capacities (stasis) or whether there is transformation (morphogenesis). Some organizations and societies lack the capacities to undertake any new round of innovation-design requiring a transformation of the existing knowledge and repertoire. Also, the development of theory and research on the diffusion and adoption of a wide range of organizational and technological innovations indicates that the pre-existing situation and its causal dynamics is central to the possibility of transformation.

The discussion of recursiveness and repertoires of recurrent action patterns can now be connected to the distinction between the re-production of the pre-existing configuration, or its transition into a different configuration as in Figure 4.1. We assume that in a process perspective chronic recursiveness is prevalent, yet contingent and varies in durability through successive cycles. It follows that configurational dynamics has two distinct and related meanings:

- As the sequences of recurrent action patterns and their durational dimensions
- The difference between the re-production of the existing configuration or its transformation.

Therefore, any alteration to the configuration is a case of non-reproduction and of morphogenesis irrespective of the 'directions' of the changes. Moreover, every claim about change must show that the pre-existing repertoire has been altered and therefore cannot be activated. The problem of measuring and confirming that change has taken place can now be formulated in terms of a recognizable alteration within either the organization or the context or the co-evolution between organization and context.

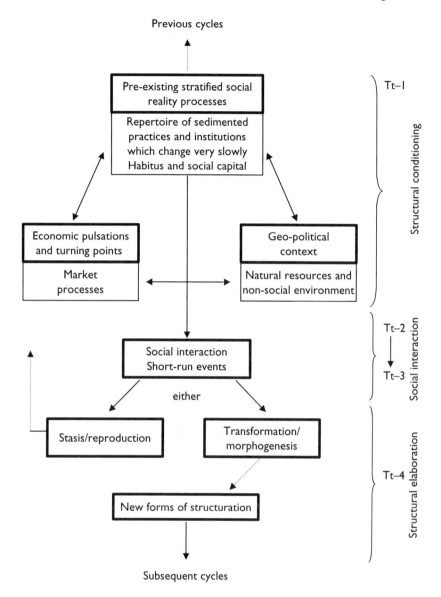

FIGURE 4.1   Long- and short-term framework: stasis and transitions (modified from: Peter Clark 2000, Figure 5.1)

In order to distinguish between stasis and morphogenesis, analysis must specify the antecedent, pre-existing cycles of structuring in the context-and-organization (see Figure 3.3). The examination of antecedents goes further than the complexity theory of starting points. The analysis of the pre-existing structural cycles in the organization-and-context also provides a frame of reference against which to assess where change – if change/transformation has occurred – and how extensive the changes have been, or are likely to be in the future. This examination distinguishes stasis from morphogenesis. Distinctions should be made between the introduction of new capabilities and the exnovation of existing capabilities. The time span of stasis and morphogenesis should be sketched with respect to both longitudinal and event time frames. Analysis should aim to sketch emergence and unplanned change as well as being aware that unplanned order can arise and falsely seem like chaos to analysts with very narrow time frames and somewhat closed system minds. Scenario-writing is one means that might be employed (Hodgson 1999).

The structure–agency issue has preoccupied theoretical developments in the past two decades and generally the solution known as structuration has been adopted (e.g. Barley, Orlikowski). Unfortunately, structuration theory, in avoiding reification, conflates structure into agency. Therefore, the realist turn argues that they should be separated sequentially in the strategy of analytic dualism (as Chapter 3).

The formulation and resolution of the structure–agency issue through the duality of structuration is problematic. This is because the temporal compacting of structure and agency undermines the recognition of the pre-existing configuration in one part of the theorizing. Giddens effectively fuses structure and agency thereby denying an ontological status to structure. His notion of a virtual structure being instantiated by agential actions denies the pre-existence of structure and suspends temporality. His notion of instantiation renders the notion of re-production meaningless. Structure and agency are sequential in different tracts of time (see Barley 1986). Giddens is criticized for an over-active view of the agent and an understratified view of the actor that strips them of inner identities and tends to be one-dimensional. Much of structuration theory on process tends towards hyperactivity because of the prevalence of the micro-level, here-and-now time frame. We should be able to theorize process as flows, cycles and movements and not be floored by the flux (see Pugh and Hickson 1976: 1). Giddens has explicitly highlighted the time dimension, but critics observe that structuration suffers from the conflation of the temporal dimensions.

Morphogenesis[1]

What is the role of stratified social reality? Archer (1995) contends that humans find that they are often located – albeit unknowingly – in positions and roles that are part of structures holding positions in defined relationships. Even so, these structures permit individual interpretation and individual capacity to alter the structure. The *realist approach* makes clear

---

1   This section draws closely from Clark (2000). There is a very useful introduction to the debate by John Parker (2000) *Structuration*, Milton Keynes: Open University Press.

distinctions between events and conditions that occur at the same time from those happening over extended periods of time. Structure is always dependent upon agency but is emergent from structure. The causal powers and liabilities inhering in structure differentially condition agential directions to action. Therefore, structure precedes agency in a sequence whereby structure sets degrees of freedom and zones of manoeuvre that place limits upon what those occupying positions in the structure should and can do. All structures possess internal relationality between those positions that are relatively enduring and ontologically distinguished from agency. There are, therefore, contextual limits to the transition of structures, yet transitions are possible.

So, how do the socio-cultural structures emerge? We should separate the causes acting on phenomena from the possible consequences and possible mutual interactions. Emergent processes could be outcomes of feedback processes, negative and positive. We would conceptualize agency in terms of degrees of freedom, providing for the possibility of selecting courses of action. Here the feedback loops are contained in those relational properties within and between organizations that actually constitute the system.

How does the analytic duality cope with social process and temporal ordering? The temporal ordering is vitally important. Social forms consist of the connections between people – the social structure – so societies are not reducible to people. Social forms have a pre-existence that establishes their autonomy as possible objects of enquiry and they have a causal power that establishes their reality. Because social forms pre-exist social activity, they entail a transformational model of that social activity. Therefore human agency mediates the causal power of social forms. Consequently, social structure is always pre-existing, autonomous and possesses its own emergent properties. This sequence occurs in processual time, not with simultaneity, but with temporality. Therefore, structure and agency have to be viewed separately in their interplay because each possesses emergent properties and must be analysed in parallel as an analytic dualism. The outcome may be the reproduction of the pre-existing structure or its transformation.

How is emergence theorized? Archer (1995) tightens the analytic specification with the three-fold sequence of:

Structural conditioning <–> social interaction –> structural elaboration

Analytic dualism requires the practical analyst to know what reality is and to be able to explain reality in respect of any problem. The pre-existing structural conditioning *interplays* with the emergent socio-cultural interaction. Figure 3.3 and Figure 4.1 show the tri-element cycles of structuring that are previous and subsequent to the focus of the analysis between Tt–1 and Tt–4. The first two phases of Tt–1 and Tt–2/Tt–3 are separable and cannot be analysed by conflating them and/or eliding between them. The emergent property of the pre-existing structure and the actual experiences of the agents involved in activities shown as Tt–2/Tt–3 are separable, not synchronized. Yet, there is interplay between them and Tt–1, the structural conditions that pre-exist. The interplay possesses an intrinsic influence of the past even though actors are frequently unaware of the real prior events that have constituted the pre-existing structure. These social structures possess a flow that can be located in phases. Social agency mediates the causal power of the social forms and this mediatory process is implicated in the first phase of any transformation cycle. People can exercise their powers and liabilities relative to the pre-existing socio-cultural powers. The pre-existing structural and cultural emergence shape the social arenas people inhabit. There is structural conditioning which in effect gives directional guidance, yet offers degrees of interpretative freedom.

The pre-existing structures are generative mechanisms that interplay with other objects in a stratified world leading to non-predictable outcomes. There is no period that is unstructured. After Tt–2/Tt–3 of social interaction in the cycle that is the subject of analysis, then the outcome at Tt–4 can be reproduction or transformation. This variant of social realism is compatible with a range of social theories. There is a long-term flow through the three steps so the steps must be split into episodes. The realist analysis aims to make statements about structuring without reference to the agent. Social structures (Tt–1) and agency (Tt–2/Tt–3) are separable. Agency only reproduces or transforms structure in any succession of processes in configurations. The distinction between structure and the agents rests upon the distinction between system integration and social integration introduced by Lockwood (1964). This permits various combinations of the two dimensions. There is a crucial distinction between the emergent properties and the relatively enduring pattern of social life. Structurally emergent properties are differentiated by a primary dependence upon material resources such as land, food and weapons. The structurally emergent properties are not reducible to people (see Elias 1994).

The analytic dualism has considerable consequences for organizational innovations. Analysts should adopt the three steps already identified:

(1) Examine the pre-existing structure and culture as an object which conditions action patterns and supplies agents with strategic directional guidance This point articulates with my notion of structure and activation by agents because pre-existing cultural systems already contain specific doctrines and knowledge (Clark 2000). These pre-existing doctrines and knowledge shape the social environment to be inhabited so that the outcomes of previous actions are deposited in the unfolding situation. These past scripts set the stage. People are distributed through the structure with limited and contingent degrees of freedom (i.e. limited choice) and they are for the moment endowed with different vested interests. Moreover, resource distribution at Tt–2 of material, symbolic and cultural resources is relatively durable. The degrees of interpretative freedom are somewhat finite. Even so, the strongest constraints never fundamentally determine the agent.

(2) Agents interpret the pre-existing in terms of their projects and the conditions. It may be that new cycle actions or some part of the cycle introduces new external conditions (e.g. Weick 2001).

(3) Archer (1995) contends that transformation arises in the third phase when certain elements are present in the second phase. Stasis is most likely to occur when there is high social integration combined with both high and low system integration. Transformation is more likely when low social integration is combined with either high or low system integration. It is the interface between the structures of resource distribution and the vested interest groups which mediates the shift from stasis to transformation. We illustrate this in Chapter 10 with examples from global media sport. The final phase is social elaboration and this arises from the previous socio-cultural interaction that was conditioned in an earlier context.

There is no intention to predict, but to pinpoint the processes guiding action in a particular direction and to explain the specific configuration that arises. This is an analytically structured narrative providing analytic histories of emergence. It is here that scenario-writing could play a considerable role in organizational action.

Technology: social made durable and relational materiality

Social made durable

The three research programmes on innovation examined in Chapter 2 treated technology in three different ways. First, as hardware objects with an autonomous life; effects were inscribed and immutable. The social was appended. Second, through the critique of technological determinism. This led to the duality of social and technological artifacts examined in time–space free frameworks occasionally within the structuration framework. Orlikowski and Yates (2000) explored the temporal dynamics of the duality framework. More generally, there was very slight unpacking of the construction of technologies (e.g. CT scanners) or of the unfolding into new structural repertoires in the longer term (e.g. Barley 1986). Temporal conflation was the pattern. The critical political economy of technology was largely ignored. Third, the socially constructed 'design' approaches transformed technology into the social made durable. We shall now examine this third research programme.

How is technology conceptualized as 'the social made durable' and 'relational materiality'?[2] The reasoning is that everyday life is dominated

---

2 This section selectively draws from the hub of scholars assembled by Callon, Latour, Woolgar, Law, Bijker, Hughes. These first-generation scholars have many connected nodes and a relational configuration containing considerable variety and no little internal debate.

by science and by the material world, containing both nature and the artificial design of technology artifacts. Our relational configurations are permeated by thousands of non-humans: telephones, climate, houses, roads, germs, earthquakes, sleeping beds, and so on. We do not live in a homogeneous, but rather a heterogeneous context. Cathedrals and laboratories are obvious illustrations, each with socially constructed features. Ensembles of multiple technologies directly affect what we do, how we work and how we live (Gille 1978). Technology is therefore context-dependent and society and technology are simply alternative sides of the same coin. Consequently, there is a seamless web between the social artifacts and technology artifacts and between them and everything else.

Examples of the social made durable include the door-groom (see Latour 1991) and the swimming pool robot cleaner (Clark 2000). First, the door-groom, high on the inside of the door of a room, closes the door automatically so that the door becomes part of the non-human wall enclosing the room. The groom does not require a full-time liveried servant to open the door when people leave and enter the cube-like enclosure of the room. The groom replaces the need for social boundaries and separates action outside the room from action inside the room. Second, the swimming pool robot cleaner occupies a pool space designed to allow the robot to criss-cross the pool in a predetermined pattern and pace so that its tentacles sweep detritus to the exit channels. The robot can be set to operate when most convenient for its users and can be adjusted to cope with seasonal and other variations in the amount of debris. Owners of the robot do need access to periodic and contingent maintenance from contractors, yet can manage without employing staff or themselves in cleaning. The robot behaves as though a script has been inscribed into its functioning by authors (actors) at the suppliers who designed the system. The behaviour of the machine as a non-human involves operating within an artificial context of the pool and with various heterogeneous actors such as human swimmers, dead flora and insects, natural elements. The robot occupies an important role in the life of its users, especially if they find human labour expensive or unavailable. Failure (e.g. break down) by the robot is unpleasant.

The groom and the robot are artificial constructions – as is all technology – and therefore represent the social made durable. Its durability as a network actant (non-human) has consequences for the social. When we initially look at a groom or the robot we rarely ask about their social origins or their social roles. This approach is at the centre of the third research programme on technology. Moreover, the 'technological frame' in everyday life shapes the activities of actors and relevant groups and is a contingent process shaped by them (Bijker 1995). Controversy and struggle among the heterogeneous network of actors is endemic, specifically over the dramas of the construction and stabilization of networks. The theorizing of power is both central and also contains many features initially illuminated by Foucault (1972, 1977). While Foucault focused upon the previously unexamined role of discourse, the school of relational materiality extends the scope to examine human agency (e.g. Pasteur and germs), attempting to assemble heterogeneous networks whose durability is contingent.

The social construction of technology systems

The social construction of technology systems (SCOTS[3]) began as an intentionally stripping away of the silent social origins of technology artifacts as indicated in the ground-breaking collection edited by Bijker, Pinch and the historian, Hughes (1987). With a rich variety of examples they contend that artifacts are the unintended outcomes of intended actions by a plurality of very diverse relevant groups.

Bijker's (1995) analytically structured narrative of the social evolution of bicycles in nineteenth-century France and England reveals that the relational configuration of relevant groups includes:

- Young males wanting a dangerous, speedy machine
- Old men wanting a safe machine
- Ladies wondering how to adapt their clothing to the exigencies of two-wheels
- Tyre manufacturers, distributors, engineers with visions (e.g. Starley), and so on.

The emerging shape and usage of the bicycle is an outcome of the conjunction between suppliers and users (Cowan 1987). The users can choose from the variety of bicycles on offer. Figure 4.2 is a rough-cut summary of the relevant groups, their definition of problems and solutions plus the emergent solutions in the 'trajectory' from the pre-existing Penny Farthing to Lawson's bicyclette. These are shown in a relational configuration that has been abstracted from the wider context. The relevant social groups articulate problems and solutions. Solutions might arise from other, different milieux. One example is the entrance of Starley from an English sewing machine firm in Coventry into bicycle production. The different relevant groups define problems according to their meanings and experience. There are many possible solutions: interpretative flexibility. Figure 4.2 shows that in the case of the safety problem there is an array of possible solutions each with new problems. So, the non-linear transformation from the Penny Farthing is an example of a relational configuration which, for a period, achieved stability because problem-solutions were black-boxed to remove the strategic level from examination. Black-boxing removes controversy among the relevant groups. In this perspective the

---

3   My preference is for SCOTS rather than SCOT. My reasoning is that the 'S' stands for systems and that is an important step towards multi-level relational configurations.

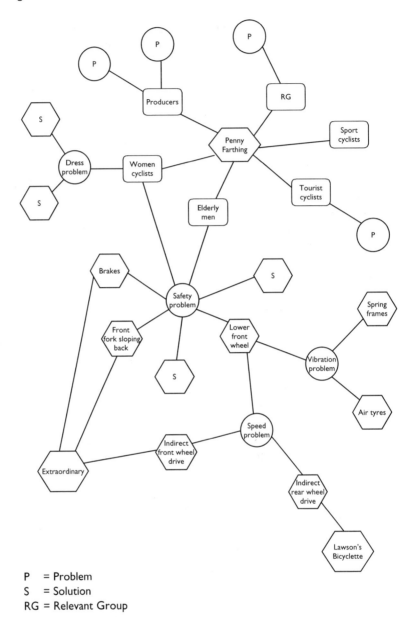

P   = Problem
S   = Solution
RG = Relevant Group

FIGURE 4.2   Relevant groups, problems and solutions (from: W.E. Bijker 1995)

powers and liabilities of the stabilized network determine the way the
technology functions until that particular black-boxing and its stabilization
of the network is overturned. There are no fixed, essential properties, but
there is interpretative flexibility.

Until quite recently the SCOTS approach concentrated upon the 'design-
and-translation' of social visions into durable forms. This was temporal

conflation downwards. They initially omitted the implications and were excessively agnostic about the political differences between positive and critical political economy. Now it is accepted that consequences should be examined. However, there has been an over-strong presumption that consequences are inscribed in scripts and that users have very limited zones of manoeuvre. Designers may seem to inscribe a vision of use, but users may start from a completely different agenda. That situation has arisen with strategic co-ordination systems (see Chapter 2). Therefore the attention to upward conflation omits central conflation and does not sufficiently achieve the required full temporal sweep.

SCOTS has shown slight interest in comparing national cultural repertoires (see Chapter 5). Although Hughes (1983) demonstrates that the electrical and social power networks constructed in Chicago and New York (1880–1920) differed from those in London, England, his analysis is relatively silent on the role of national innovation design systems.

Social capital hypothesis

This section reviews the social capital hypothesis and its implications for process innovations. The social capital hypothesis was deployed by Putnam (1993) to explain the outcomes of the decision of the Italian state to decentralize from Rome to the regions through legislation activated in 1970. The decentralization legislation was a common process from a legislative perspective yet was located in quite varied local processes of the pre-existing political systems. Putnam concluded that the actual form of decentralization that emerged contained two quite different capabilities. One form was typified by cabinet stability, budget promptness, provision of information services, reform legislation, innovative legislation, day care centres, family clinics, industrial policy instruments, agricultural spending capacity, high local health expenditures, housing and urban developments, bureaucratic responsiveness. The other was typified by the absence of these capabilities (Putnam 1993: 12, 67f, 75). Putnam and his colleagues skilfully applied proxy variables to reconstruct the interaction between: pre-existing processes and state-initiated innovation. Those differences were best explained by the pre-existing, historical political repertoires. The content of the civic culture was a key variable explaining the activity of democracy. The implication is that the social capital of some Italian regions explained the presence and absence of strong and weak capabilities in decentralized government. The virtuous regions possessed a civic culture with norms of generalized reciprocity and trust and these norms facilitated co-ordination and reduced the incentives for opportunism. Their repertoire is a cultural template for future collaboration.

Social capital can take other forms than simple democracy. There is the case of extra-legal actions. De Soto (2000) formulates and explains the mystery of financial capital through an examination of another facet of

American social capital. De Soto asks 'why does capitalism triumph in the west, especially America, and fail everywhere else?'. He asks the question because with the advantage of hindsight he is clear that the modern developmental approach to economic development formulated in the 1950s and 1960s worked much more effectively in the west (e.g. Europe: Djelic 1998; Kipping and Bjarnar 1998) than in the rest of world. De Soto concludes that the difference rests on how *extra-legal claims to property* were resolved in the USA since the 1860s. That process explains the success of capitalism and therefore the absence of similar solutions outside the west is not understood. This is the social capital of finance capital.

Why does capitalism thrive so well in the west? De Soto contends that the issue is the ability to produce capital from savings such as the buildings occupied by families. The key is whether resources are dead capital whose ownership rights are inadequately recorded. Only recorded rights can be transformed into finance capital by using concrete capital (e.g. buildings, land) as collateral. In the west, every parcel of land is represented in a property document that is part of a state-enforced system. Outside the west, these forms of concrete capital (e.g. buildings) lack the processes necessary to create finance capital. Most of the world's concrete capital is dead capital. Capital, as a resource detached from concrete entities, means:

- Being able to fix the economic potential of assets separate from their physical state
- Integrating dispersed information about concrete assets into one system
- Making people accountable for their property
- Making assets fungible so they can be divided, combined and mobilized in business activities to suit any transaction
- Converting citizens into networks. The formal property process enables the construction of an infrastructure of connecting devices such as transport systems in which buildings are nodes on a rail system
- Protecting transactions in public record keeping and through state-based legal policing.

Consequently, in the west the house can be used as an asset to generate loans for various activities. The issue is whether this western 'bell jar' (Braudel 1982: 248) can be opened out. Can non-western nations create similar formal property systems? That would require international agencies to exnovate (take-out) much of the simply economic analysis of financial capital.

De Soto's analysis prises open the hidden role of extra-legality as illustrated in the epic Hollywood films of the nineteenth-century American land claims. All through Europe, from the late sixteenth century onwards, there were extra-legal claims to lands based on their occupation and their use by non-owners. Frequently, these groups formed local associations to protect their claims. Some European nations quickly regulated the extra-legal claims, thereby altering the basis of commerce. This recent process is less than three centuries old and its unfolding in the USA is even more recent. In the USA there were extra-legal claims backed by local associations for their protection and eventually the state supported those claims based on the usage of land. De Soto maintains that this solution enabled financial capitalism.

The theory of social capital has been fully developed by Coleman (1990) and by Bourdieu (1977), yet Putnam's analytic focus is still relevant for his interrogation of why Americans no longer go bowling together (Putnam 1999). We know from public statistics and research surveys over the past four decades that American civic engagement with activities such as the local church and the Parent Teachers' Association has declined. Also there has been great decline in bowling together and taking part in leagues. Putnam takes the view that these trends are durable and consequential because activities like bowling created local networks and those networks could be mobilized to undertake actions central to process innovations.

Putnam's view of social capital is graciously romantic (see De Soto, Bourdieu, Coleman) and certainly enrols a constituency through its political network. However, we have every analytical reason to expect that the shape and form of social capital would change after the mid-nineteenth century, and the more recent development of capacities for mass customization suggest greater concerns with individual quality time. Certainly, the growth of the global service class complicates the loci of power. Therefore Putnam might examine events that probe the actual do-ing of social capital in a network society. For example, the recent 11 September demolition of the twin towers in Manhattan provided many examples of actions taken that represent a modulated, differentiated, articulated and collective response activating of social capital in support of an infrastructure of public and local expert systems (see Putnam 2000).

Circuits of power

One problem is how to construct a useful analysis of the circuits of power involved in innovation for multiple levels of analysis: the state, professions, firms, localities and individuals. Clegg (1989: 211–240) provides a suggestive framework that draws on Callon and Latour. Clegg incorporates both agency and the distinction between system integration and social integration to create three levels of circuit each typified by a distinctive form of power. This is shown in the three left-hand columns of Figure 4.3. The dynamics are shown in the relations between the eight elements on the right-hand side. The framework highlights the contingent dynamics in which innovation unfolds and highlights the role of organizational innovations.

Innovation is frequently treated as an episodic event in which some agency (e.g. a firm) masters an existing situation and successfully introduces a new innovation that brings in more satisfaction and greater profits. Frequently the agency is an individual. However, as we now appreciate, agency is located in a pre-existing system of social relations as shown in

Circuits of power

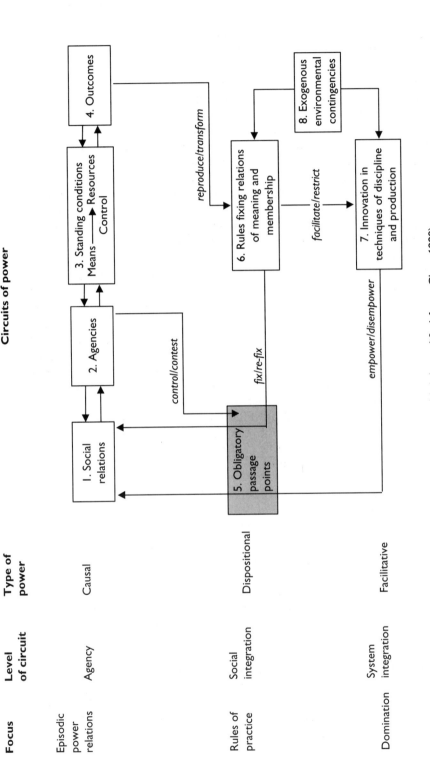

FIGURE 4.3   Circuits of power (slightly modified from: Clegg 1989)

box [1] and box [2] of Figure 4.3. Headed arrows connect these elements. The episodic power circuit of agency is at the causal level and so they encounter standing conditions [3] from which there are outcomes [4]. The key issue so far is whether the agencies possess control or not and whether the outcome reproduces or transforms the pre-existing social and system integration. For example, in the United Kingdom the Conservative state sought, after 1981, to introduce market-related innovations into the welfare capitalism of the public sector (e.g. hospitals) in order to accelerate the rate of innovation in care and to reduce the impending fiscal crisis of the state. The struggle between the Conservative state and the powerful medical profession is one of contest – mainly covert contest. Despite the introduction of a market mechanism in the flow of patients from primary care into the hospitals, there was considerable reproduction of pre-existing social relations. By comparison, the state was able to achieve slightly more innovation in the areas of education and in the prison service. The framework of circuits of power shows that the circuit of agency has to engage with the circuits of social and system integration (see Archer 1995). The circuit of social integration is that of dispositional power embedded in rules of practice. These rules of practice fix (and re-fix) the relations of meaning and the rules of membership of particular strata. Thus any attempt at innovation by the state encounters dispositional power and this may facilitate or restrict the attempt at innovation. In the case of restriction, the circuit flows into the heartland of the 'obligatory passage points' (box [5]) in the circuits of power and is rejected/curtailed. In the case of facilitation at the dispositional level, the circuit reformulates dispositional power and alters the rules of practice. The third circuit of facilitative power and domination also flows through the obligatory passage points. The circuit of system integration involves facilitative power and requires domination by one agency over other, rival agencies to bring about innovation. Otherwise stasis is reproduced. Figure 4.3 also locates exogenous environmental contingency (box [8]).

The framework of circuits of power provides a useful interpretation of Weick's (2001) widely admired model of social evolution and organizational sense-making, but introduces some difficulties in Aldrich's (1999) stimulating account of organizations evolving.

The circuits of power framework highlights the much-neglected role of rules of practice (dispositional power) and domination (facilitative power). The attention to obligatory passage points, through which attempts at innovation must travel, is a useful corrective to the strategic fantasies of much of the innovation literature. Equally, the circuits of power framework allows for emergent, unintended outcomes. The seminal example of unintended outcomes is the seventeenth-century struggle between three agencies – the English monarchy, the landed aristocracy and the London burghers – from which emerged the political system of the House of Commons and the House of Lords.

One requirement of this analytic framework is that innovation requires a longitudinal approach in which power is analysed as both strategic and also contingent upon relationships between different agencies. A central issue is the making of alliances both between humans and non-humans.

Clegg contends that in late modernity the three key nodal points of power were the state, the market and corporations, and they were counter-balancing one another. However, in the new political economy the market has emerged as the dominant circuit and provides the architectonic around which the state and corporations become articulated (1989: 273–275). The role of the market exemplifies the plural and diverse pathways through which political traffic passes (e.g. innovation). With the market there is an emphasis upon seduction rather than overt and covert repression, and therefore a shift to dispositional and facilitative circuits of power from the episodic agency. In the new world of mass customization, the paramount tool of system integration to continue the reproduction of domination becomes 'seduction'. The pursuit of brands and things becomes an encompassing passion. Therefore the market becomes an obligatory passage point through which a configurational field is stabilized. The market is also a black-hole sucking in agency and outputting diffuse power. Consequently, dispositional power rather than the traditional spectacle of sovereign power occupies centre stage. Clegg suggests that there is 'a post-modern democratic freedom of the market' (1989: 274). This seems to be a serious rather than a playful observation and contrasts with Harvey's (1989) view that some corporations may be able to shape their markets. So, although Clegg engages with the world systems perspective of Wallerstein, it may well be that the exceptional position of American multinational firms is underplayed (Clark 2000). Of course, even these firms also face the market.

Global Contexts and National Innovation-Design 5

This chapter connects three lines of analysis in an eclectic theory. First, the new positive political economy of global competition between contexts is being developed. This theme is heavily dominated by economics and geography and tends to be strong on space and thin on geo-history. We scrutinize the controversial and highly cited framework developed by Porter (1990, 1998), but revised by Dunning (1993) and Rugman (1995). The revised framework can play a useful role in directing analytic attention.

Second we explore national cultural repertoires, institutions and the isomorphism hypothesis, isolating mechanisms and elective affinities. There is a strong influence from cross-cultural management (see Chapter 3) and therefore both the historical and the placeness require considerable development. Cultural repertoires contain isolating mechanisms and are significantly formed through the market. Many research programmes examine the diversity of capitalism with a relatively open mind about convergence around a singular form and with increasing scepticism about the notion of the USA as the only leading-edge context. The concept of national institutions is being replaced by the concept of cultural repertoires and by recognizing that there *are significant intra-national variations*. The role of national markets has become central as milieux of learning by firms (see Chapter 6). Within those milieux the service class of professionals, scientists and managers occupies a key role. Consequently, a similar genre of organizational innovation is constituted differently in each national context. There are many carefully researched examples of these cross-cultural variations, ranging from long-established forms (e.g. Scientific Management) to Lean Production and Enterprise Resource Planning Systems. These contain core symbolic features surrounded by considerable national differences. There is an elective affinity between the cultural repertoire of the nation and the specific variation of organizational innovation found in the nation. The area of corporate co-ordination is highly sensitive to national predispositions as interpreted by the domestic service class. British-owned firms are prey to the 'possessive individualism' that occupies a significant tendency in the national culture and explains such details as queuing systems in hospitals.

Third, we explore the role of global firms. The economy and cultural repertoire of the nation is significant in shaping the zones of manoeuvre and proximal development for its domestically-owned firms. However, firms may leverage considerable autonomy, especially firms from smaller nations that seek to gain their advantage by being based in several nations (see Porter 1990).

Global context and national clusters

This section presents a focused synthesis drawing from the debate stimulated by Porter's (1990, 1998) analytic framework for explaining the role of nations, sectors and firms in national innovation-design systems. Porter's analytic framework (the diamond) and the empirical studies have been both highly cited and robustly criticized (e.g. Dunning 1993; Davies and Ellis 2000). In addition to these analyses, we shall argue that Porter's approach largely excludes the analysis of the cultural repertoire and is essentially time-bound. However, Porter has succeeded in both surfing and stimulating relevant analytic developments. Therefore a synthesis of Porter, Dunning, Rugman and others is formulated in Figure 5.1 (Clark 2000: 140–154).

Dunning (1993: 98–101) developed an eclectic, location-oriented framework to inform macro policy-making about the dynamic interplay between multinational businesses and the competitive advantage of location in nations. He contends that location is a major cluster of variables and that within the global economy there is competition between locations. Dunning underlines the key position within any nation of foreign multinationals. For example, the entrance into Europe of Wal-Mart and its relation with key European firms such as Unilever has considerable implications for the UK, France and Germany. Figure 5.1 contains the influence of Dunning in the presence of *foreign multinationals* [7] inserted into Porter's original framework.

Porter (1990, 1998) contends that the three digit sector (e.g. wines rather than food and drink) is a better unit than the firm for analysing the competitive advantage of firms and nations. Sectors are sometimes agglomerated into regions (e.g. wine growing in the Napa Valley of California) and different complementary sectors cluster together. Most nations only possess a competitive advantage in a small range of sectors and only a very few nations possess an advantage in many sectors (e.g. USA, Germany).

There has been much critique of Porter's data and the linkage to the framework,[1] but the usefulness of the sector as a unit of analysis has, for varied reasons, proved sufficiently robust (see Clark and Mueller 1996).

---

1 I find Porter's contribution pragmatically useful for its contribution to reformulating the previous agenda of problems. I also have a degree of scepticism about the neatness of fit between the case studies and the analytic framework. My chapter seeks to show that the critique needs to be more substantial (Clark 2000: Ch. 7) and to involve the revisions I have set out in this book.

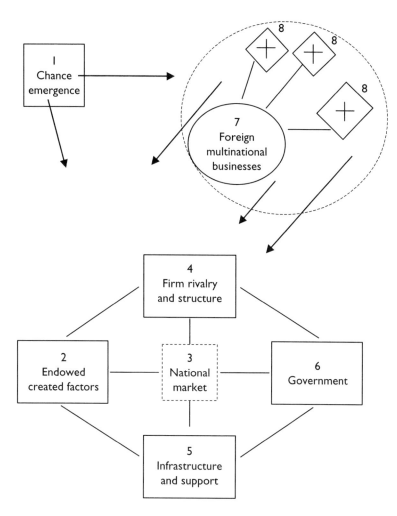

FIGURE 5.1   Eclectic theory of competition between contexts
(from: Peter Clark 2000, Figure 7.3)

Porter's contribution to Figure 5.1 is in the six factors listed 1–6. These factors are an inter-acting, dynamic configuration. The diagnostic logic is that for a sector (e.g. pharmaceuticals) to succeed in the world political economy the domestic context should contain world-class performances in each of the factors. Therefore, explaining Italian success in the tile sector depends on sustaining the whole framework. The role of *chance* and the pervasiveness of *indeterminacy* [1] about the future are important. My revision places the domestic and elected *market and demand* conditions [3] at the centre. Market characteristics in the form of demand provide the opportunity and impetus to firms and shape the zone of learning (Penrose 1959; Clark 2000). For example, in consumer goods from television sets to cameras, the Japanese consumer has provided a large, demanding, rapidly saturated market (Porter 1990). The role of the market is examined further in Chapter 6.

Factor conditions [2] contains two rather different elements: non-human endowed factors and created factors, including knowledge. Porter treats the endowed factor only as objective, thereby limiting his account. Endowed factors should also be theorized as socially realized. First, *endowed factors* are largely those of geo-political position, terrain, climate and resources. Porter contends that these are less important despite their not inconsiderable role in the itemizing of American sectoral strengths, as in exporting rice. For Porter it is the USA and Japan that provide the most instructive comparison and contrast. The USA is a large area containing some of the world's finest real estate: the North Atlantic and Pacific coast lines; rich internal river ways and lakes for internal navigation and electricity power; varied soils and forests (hard and soft woods) with extensive fauna; a varied, and mainly beneficial, set of climatic conditions; available resources of coal, gold, oil, water for hydro-electric power and other natural resources. Japan is 2,000 mountainous islands in an earth-tremor zone off the coastline of one of the world's most significant political systems: China. Porter, writing in 1990 (see Porter et al. 2000), concluded that Japanese sectoral performance was achieved despite a poor geo-political endowment. This leads on to Porter's pre-occupation with knowledge. Second, *created factors* include knowledge plus their embodiment in infrastructural transport systems (e.g. roads, canals, ports and railways). Porter primarily examines knowledge through the composition and performance of the education system and through the role of centres of advanced knowledge agglomeration and creation. Porter contends that Japan is typified by strong knowledge distribution in basic education and by the location of research, development and design within the corporations rather than within advanced centres in higher education. It is in this latter area that the specialized university and research degrees of Germany and the USA possess common tendencies, yet very different institutional arrangements. Porter emphasizes the role of the specialist research universities: medicine at Johns Hopkins; wine-making at the University of California, Sacramento; information systems at MIT. Also there is a major role for connected, yet independent advanced research centres benefiting from the federal framework of investment. Knowledge agglomeration, creation, commodification and diffusion in the USA are remarkable. Porter does not sufficiently consider geo-political position (Clark 1987, 2000).

Firm rivalry and structure [4] allows Porter to stress that the greater the competition, then the more firms will be pressured to upgrade their capabilities and therefore become more competitive in the global economy. This somewhat revises his earlier injunctions in the five forces or the value chain. *Related and supporting industries* [5] refers to the clustering around key firms of knowledgeable, world-class suppliers, covering design, advertising, components, distribution and so on. The role of *government* [6] is to

reduce the expenditure by the state and to regulate the economic environment to create competition rather than protection. Porter's injunctions for
small government and a large 'free' market captures the spirit of the
positivist political economy research programme for the period. Porter fails
to acknowledge the central position of the American federal state in the
Cold War-related markets that stimulated electronic networks (Castells
1996, 2001).

Figure 5.1 is useful for its depiction of relevant factors and their
interconnections, and the attention to sectoral clusters acknowledges the
role of location, place and space. However, the simplifications sometimes
imposed by structured methodologies create a number of problems:

(1) The world political economy is not a congeries of geographical
     contexts like Silicon Valley and the Ganges Basin. It is a stratified,
     political reality with an uneven distribution of economic power in
     which the members of G8 seek to regulate interests and conflicts.
     There is, as Dunning rightly emphasizes, a significant domestic base
     for some global firms which is not available to all firms. In the new
     political economy some nations seek to establish their hegemony
     through collective trading agreements (e.g. NAFTA).[2] The USA gains
     access to Canadian water supplies. These organizational innovations
     are highly significant space–time colonization.
(2) The factors are loosely defined and overlapping.
(3) The analysis of contextual clusters insufficiently exposes and antici-
     pates clusters that are or have become ineffective. In the British
     context, there has been a major clustering of food, drink and textile
     firms around key retailers, especially around Marks & Spencer (Clark
     and Starkey 1988). That cluster seemed to create a world-class firm at
     its centre, yet the firm never gained a significant, profitable market
     in the USA, Canada, Asia or continental Europe. Currently it is in
     serious difficulties even in its domestic context.
(4) It is more than likely that firms in small countries (e.g. Sweden)
     construct their diamond from a base of other nations. Volvo (after
     1927) aimed to build markets in Sweden, The Netherlands and the
     USA. Firms from the large nations of the USA, Japan and Germany
     can probably use the diamond internally. However, Benetton from
     Italy certainly benefited from activity in neighbouring nations (e.g.
     Switzerland, Austria and Southern Germany). Moreover, as Dunning
     (1993) observes, European firms such as Nestlé, SKF, Philips, Unilever
     and BAT draw upon several national bases.
(5) The analysis of organizational, institutional and societal factors is raw
     and requires complementary developments. The application of this
     multinational framework requires rich analytically structured narra-
     tives.

---

2 Alan Rugman has made this point very incisively.

Cross-cultural: Child

Child's (2000) evolutionary framework for cross-national analysis aims to synthesize the Hofstede approach and more institutional approaches. Child positions the international and national levels and the interaction with the world of firms. This is shown in Figure 5.2. The non-corporate world is divided into three systems: the international and national 'material systems' (economic and technological) and 'systems of ideas' (cultural values and norms of behaviour) plus the national institutions that are regulatory, competence-forming and provide services. These three systems interact with specific firms.

Firms are conceptualized as possessing structures and processes. The role of strategic choice is to design the organization and action feedback on organizational performance. The firms' strategic choice and organizational action faces both task contingencies from the material systems and value preferences from the systems of ideas. Strategic choices are in a population economy framework facing de-selection. Therefore the level of the organization's performance mediates the task contingencies and value preferences. Firms can therefore, potentially at least, gain a zone of manoeuvre.

The institutions impinge on firms because their contextual capabilities are potential constraints on strategic choice by the firm. The firm counters these institutional constraints and capabilities by negotiating support and terms with various institutional bodies.

Isomorphism hypothesis[3]

Institutional agents: the state and professions

The equal power balance between the state, the market and corporations has been altered in the new political economy (Clegg 1989). The market has become influential. The new institutionalists maintain that the market is more properly theorized as a set of conventions than as a collective actor or a social agent (Scott 1995: 113). DiMaggio and Powell contend that the foremost engine of rationalization is the state and the professions. This section examines their account of the isomorphism hypothesis.

Earlier approaches to organizational innovation detached the firm from its institutional, societal and international contexts. The aim of the new institutional theories is to re-focalize the firm in the societal infrastructures that support and constrain firms (DiMaggio and Powell 1991). Scott (1995) appropriates structuration theory, adopting the three pillars (regulative, normative and cognitive), each of which elicits different and related sources of legitimacy. First, the basis of the *regulative* pillar is conformity

---

3   This section extends Clark (2000) and benefits from discussions with Candy Jones (Boston College).

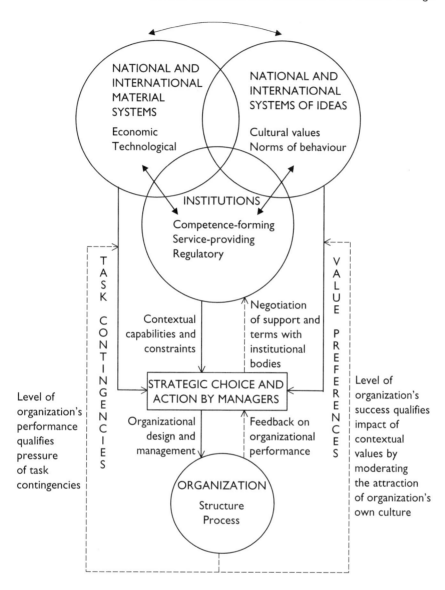

FIGURE 5.2   Evolutionary framework: cross-national organizational analysis (from: John Child 2000)

to rules. Firms have to conform to a context of rules (e.g. probity of public information about profits). Regulations are embedded in systems of domination, power, recurrent action patterns and standard procedures. Regulation establishes laws and rules, applies rewards/sanctions and reviews the extent to which rules are accepted or rejected. The state occupies a key position in enforcement. For Scott, unlike Giddens, the regulative pillar is strongly influenced by economic history (e.g. North 1990) and by transaction costs (e.g. Williamson 1985). This gives a strong tone of actors' instrumental, material interests in a cost framework that includes the issue of enforcement. Second, the *normative* pillar is the basis of compliance and feelings of social obligation. Norms are embedded in the authority system and in the routines of conformity. Normative systems connect values to preferences and what should be done in a logic of appropriateness for moral and ethical evaluations. Norms define goals and socially approved means as well as the content and actions expected of roles and positions. Therefore actors are influenced by these logics in a social situation rather than their instrumental benefits. The normative pillar was used very heavily in the modern frameworks to conceptualize knowledge, science and technology, but has been used very sparingly since, particularly in North American studies. Third, the *cognitive* pillar has been and is the current focus of attention (Mizruchi and Fein 1999). Cognitive refers to the onto-logical rules about reality and the construction of classificatory systems that frame the actors' meanings. The source of legitimacy is correctness in relation to conceptual frameworks. The cognitive is embedded in scripts. The cognitive rules define the categories to be used and how typifications are constructed.

Scott's account of the roles of the state contrasts with the account offered by Porter to include defence, the monopoly of violence through legitimate coercion and the enforcement of property rights by setting the conditions of private ownership. The state is a collective actor, acting semi-autonomously through regulative processes, and is a distinctive configura-tion of institutions. States can provide different arenas within which conflicts between organizations can be adjudicated (Scott 1995: 95). The state regulates the role of the professional associations that have control over the application of certain types of knowledge (e.g. medical knowl-edge). The professional associations occupy a crucial role in both the control of segments of societal action (e.g. absence from work) and also in shaping action within firms. The accountancy professions significantly control the audit of corporate performance in every type of firm and have utilized that role to gain a strategic position in advising firms on how to use the new information technologies. The major accountancy firms are leading proponents of the claim that knowledge is capital. Also their capabilities in auditing performance have been utilized by states in varying ways. Power (1994) argues that Britain has become an audit society with the influence of the system of audit dispersed through the whole public sector, thereby creating a transparency in performance. In Britain, users of the educational and medical systems can now consult extensive league tables which even detail the performance of local, specific schools and of hospital departments.

Before examining the three mechanisms of isomorphism, it should be noted that there is the problematic issue of the linkages between the three pillars and their role in the shaping of organizational innovation in particular societies. Each of the three pillars seems self-contained rather than generating the analytic scope of Giddens (1984) in his theory of signification, domination and legitimization. The regulative pillar tends to make compliance overpowering and understates the dynamics and variability of actions. The cognitive pillar emphasizes ideal and cultural formulations, coupled with mimetic isomorphism. This neglects the longer-term historical narrative while also confining the influence of future-oriented policy studies. The three pillars of the typology offer different not similar interpretations of the social realism versus social construction debate. Although the connecting of the normative and cognitive pillars constrains rational actor theorizing, the cognitive pillar is incomplete.

The three pillars framework addresses limitations in the Dunning–Porter framework, yet fails to build upon middle-range realist theory (Hirsch 1997). The three pillars exclude the wider social structure, suppress conflict and direct attention away from emergent transitions. Also omitted is the capacity of large, powerful multinational firms, including the international financial firms, to shape their context and to impose definitions on domestic populations (see Dunning 1993).

Three mechanisms of isomorphism: DiMaggio and Powell

The isomorphism hypothesis (DiMaggio and Powell 1983, 1991) contends that organizational fields tend to homogenize organizations through three specific mechanisms embedded in the three pillars to create isomorphism among organizations. DiMaggio and Powell focus analytic attention outside the firm on the institutional context and those institutional mechanisms through which firms might be induced, coerced and regulated to adopt structures and cultures that do not necessarily enhance economic performance. There is a propensity within fields of organizations that are interrelated to possess similar patterns of structure, action and outputs. This is institutional isomorphism and occurs because organizations seek legitimacy from the state, from professional associations and from other key organizations on which they are dependent. Legitimacy is granted only if organizations conform to certain requirements. Once innovations achieve legitimacy in a field then many organizations accept their validity and adopt the innovation.

Three themes are emphasized. First, the rejection of the orthodox conceptualization of the environment–organization relationship in organizational analysis is central. The isolated firm is re-focalized into the societal institutions where the state is a major source of rules and legitimacy. Market processes are shown to be more like networks of rules with a strong institutional signature. The state often plays a key role in the position of sectors as part of national priorities (e.g. Japan). There is a society-wide system of inter-organizational networks with cultures, roles and rules that are nationally specific. Organizational structures are externally authored, imposed, imprinted and induced. This embedding of organizations affects the content and form of innovations imported from outside. So, when American college males in the 1870s imported association football and rugby union, the emergent outcome became American Football (Clark, 1987). Consequently, the importation by American organizations of Japanese innovations might also be anticipated to lead to hybrid and novel outcomes.

Second, the new institutionalism locates the firm in a wide array of inter-organizational networks and national propensities. The symbolic role of the formal structure is highlighted. Organizational fields are typified by slow alteration and inertia is explained by the imperative of maintaining legitimacy. Cognitions are primarily explored through routines, scripts, classifications and schemas. The firms' strategic agenda is disciplining the members rather than the policy relevance. Conflicts and struggles are located more in the organizational field than inside the organization.

Third, institutions are theorized as stable templates (designs) for sequences of activity that are recursively reproduced. The templates and their discourse provide a grammar of strategic action rooted in the past. It is the taken-for-granted elements of action, especially the classification schemes and the situated, practical consciousness – knowledge without concepts – that is embodied in the reproduction of social and organizational structures. There is regulated improvisation on the templates.

Organizations facing similar environmental conditions are becoming more homogeneous in their structure, culture and actions. The similarities arise from advanced bureaucratization. Organizations are becoming similar without becoming more efficient. The process of becoming similar is referred to as isomorphism. The economic role of the market mechanism is slight in well-established fields. Consequently, the ritual, ceremonial and political features of organizational life overwhelm the narrow pursuit of efficiency.

Organizational fields are situations where firms and organizational units face similar environmental conditions. Organizations include key suppliers, regulatory agencies, consultancies and customers. Organizational fields emerge over time and will be most homogeneous when four certain conditions apply:

(1) When organizations are very dependent (e.g. centralization of resource supply).
(2) When uncertainty is high.
(3) When firms rely most on sectoral and professional associations.
(4) When state transactions are high there will be few visible alternative templates.

Isomorphism is of two main types: competitive and institutional. Competitive isomorphism needs supplementing by institutional isomorphism. At the birth of a new field, the market mechanism is the most influential, but once the field is established the professions, other organizations and the state become pre-eminent. Institutional isomorphism is examined through three mechanisms that shape homogenization processes within specific organizational fields:

(1) *Coercion* mechanisms can range from the imposition of templates and rules (e.g. legal requirements for audits) to the equally influential, though more subtle, inducement to adopt rituals of conformity in order to gain legitimacy. Coercive and regulative processes arise from the cultural expectations in the host society and from the state (central and local), and from dependence upon powerful organizations.

(2) *Mimetic* mechanisms apply with contextual uncertainty. The sources of uncertainty are extensive. Copying enhances legitimacy. There is a relatively small pool of variations from which options can be selected. Even international firms purchase new templates from a small set of consulting firms. They claim that the history of management reform is dominated by instances of isomorphic modelling. The selection process for innovations is more influenced by isomorphic tendencies than by evidence that efficiency will improve.

(3) *Normative* isomorphism largely arises from the collective struggle of professionals to define the conditions and methods of their working and to control other parties. Professionals typically seek to establish areas of jurisdiction around a cognitive base that provides legitimacy (e.g. the audit) and occupational autonomy (e.g. accountancy). Of course professionals are subject to various coercive and mimetic pressures, yet it is the growth of their networks which provide the channels down which homogenizing tendencies can travel (e.g. adoption of computer-based integration). Professions filter and socialize their personnel, thereby enabling and privileging normative isomorphism.

Review

This research programme heightens the role of inter-organizational networks in innovation. The attention to the issue of isomorphism and to the three possible mechanisms is highly relevant. The notion of the market as a social-cultural system is developed in Chapter 6. Six amendments and revisions to the new institutionalism are underway:

(1) The new political economy situates institutions and organizational fields within multi-level frameworks extending the societal level into the world economy. This requires a more historical and comparative form of research.

(2) In the initial studies, the temporal dynamics of organizational fields were compressed into orderly linear models that highlighted the cumulative reproduction of existing patterns. However, organizational fields are the intersection for many contradictory developments arising from different sources creating indeterminacy and elements of hybrid emergence. There are conflicting systems of rules and varying forms of knowledge.

(3) Current research examines how the life course of fields contains several key moments of transition and de-institutionalization of existing fields. Research is examining the Anglo-American states and their policies of deregulation of the public and welfare sectors. Studies of the dynamics and politics reveal emergent potentials arising from contradictions, competition and conflicts. There is an important bridge to be crossed between those developments and studies of the life course of major sectors, especially with respect to the role of design hierarchies.

(4) At the organizational level there are strategic responses of varying kinds to institutional processes of isomorphism (Oliver 1991). Differentiated organizations develop specialized, protective and cosmetic capacities for dealing with the context.

(5) Intra-organizational dynamics shape the zones of proximal development and manoeuvre available to organizations within particular fields.

(6) More attention should be given to institutional change and power as well as explaining the origins, reproduction and disappearance of institutionalized forms. Patterns of inter-organizational competition, influence and co-ordination need closer examination. The autonomy of levels and the role of the central state need further unpacking.

The new institutional theory claims to possess a processual-longitudinal and comparative perspective in two elements: the historical element and change of the institution. Critics suggest that the extent of dislocation and transformation in pre-existing social fields is not easily accommodated in recent studies.

Isolating mechanisms and organizational heterogeneity

Isolating mechanisms

The new institutionalism has been criticized for overstating isomorphism. The counter-perspectives, especially the resource-based view, start out from the claim that heterogeneity among firms is the predominant feature (e.g. Nelson 1991; Clark and Mueller 1996; Miller 1997a). Structuration and institutional theory in the new political economy contend that firms are often inertial captives to their own history and of social influences of the sectoral and national contexts. Consequently, firms may make fateful, ineffective decisions about their resource capabilities and are rarely able to imitate the cultural repertoire of other firms. Therefore non-rational choices are likely. This is contrary to the resource-based view of the firm (Grant 1998). The concept of isolating mechanism refers to barriers in and around a firm that prevent other firms from imitating its capabilities, or prevent that firm from imitating other firms' capabilities.

In the resource-based approach the concept of isolating mechanism explains and prescribes why other firms are unable to undertake mimetic isomorphism. There are features of resources that prevent replication (see Grant 1998). Institutional theory suggests that firms which are influenced by their social context may be unwilling and unable to imitate other firms. We can presume that these isolating mechanisms kick-in at both the strategic level of the firm and also from the institutional context. The explanation of the resource-based approach centres on strategic barriers. So there are two forms of isolating condition:

- Firms are unable to obtain or deploy resources (strategic)
- Firms are unwilling (institutional).

Oliver (1997) contends that a firm's sustainable competitive advantage will depend on its ability to mobilize only those political and cultural supports for the use of rent-generating resources.

Isolating mechanisms are the result of the rich connections between uniqueness in the firm and its surrounding institutions and also causal ambiguity (Mahoney and Pandian 1998). Isolating mechanisms are analogous to mobility barriers at the level of the firm and to entry barriers at the level of the industry. For firms with sought-after outputs from their resource base, the isolating mechanism is a barrier to imitation that other firms have to break (see DiMaggio and Powell 1991). The concept of isolating mechanisms provides insight but is the insight generalizable? Mahoney and Pandian (1998: 372–373) list more than 40 examples of isolating mechanisms, including the following: legal restrictions on entry; economies of scale combined with imperfect capital markets; high sunk costs of investment; investments with high exit and switching costs; first-mover advantage; idiosyncratic assets; enacted complexity; ill-defined property rights, patents and cognate, co-specialized assets; reputation and image; organizational capital; unique historical conditions; embodied management skills and Penrosian learning; Schumpeter's resource combinations; invisible assets; time compresssion diseconomies; response lags; valuable heuristics and processes that are difficult to imitate; distinct corporate culture; resources with limited substitutability and valuable, non-tradable resources, etc. The extensive list illustrates the challenge that firms face when located with the analytic framework of isomorphism. Copying does not come easily, except in terms of the more cosmetic structural features of the firm and particular issues affecting firms such as greening. Corporate environmentalism is argued to be morally hollow and Crane (2000) contends that corporate greening is a form of amoralization.

The Trust versus the Moguls in the American Cinema

Candy Jones's (2000) compelling analytically structured narrative of two competing networks of American entrepreneurs in the founding era of the cinema industry integrates institutional and resource-based approaches. One group is the Trust and the other group is the Moguls. Entrepreneurial action and their social networks did shape the expectations of the consumers. Use of the realist turn focuses attention on causal mechanisms and their role in re-production and in transformation. Entrepreneurs as social agents provide capital and social resources. Institutional isolating mechanisms are a selection device that encounters entrepreneurial variations in organizing the cinema industry. The isolating mechanisms create differentiation rather than isomorphism in the competitive struggle between the Trust (1895–1931) and Moguls (1895–1931). The Trust have very strong positions in film equipment and are quite strong in film stock, production and distribution, but not in exhibition. The Moguls' strongest positions are in exhibition and distribution, with a presence in firm production. Their competing networks between 1895, 1907 and 1931 represent different learning processes based on distinctive mental models, templates and organizational building stocks.

The Trust and the Moguls possessed quite different resource bases. First, the identity and background of the Trust was mainly in invention and manufacture, plus some theatre experience, while the Moguls' background was in retail and in exhibition. Second, the macro-culture of the Trust blended French, German and American influences with a mixed American group. The Moguls were predominantly of East European Jewish extraction, with a minority of West Coast Americans. Third, the politics of control among the Trust was very heavily gounded in litigation, while among the Moguls partnership was much more significant than litigation.

The Trust and the Moguls constructed quite different strategic networks. First, the Trust's institutional network possessed a strong technical focus and became legalistic, with the electrical industry as the source of key templates. The Moguls' network possessed a consumer focus on macro-culture and deployed templates from Vaudeville. Second, in terms of capabilities and structures, the Trust began as an atomistic collection of independent firms in manufacturing and after 1908 became more centralised with governance through legal contracts. The Moguls' experience was in exhibition and distribution, and they developed marketing prowess in feature films and with brands based on stars. Third, the strategic orientation of the Trust started with patents on the technology and then, when these expired (c. 1910) and, despite expensive lawsuits, were declared illegal in 1915. The Moguls focused on attracting audiences and then on vertical integration. Fourth, the market orientation of the Trust was on the growing demand yet they found half the market occupied by French films. After 1908, as demand grew rapidly, they adapted to changes in consumer taste by moving from shorts and actualities to feature films. The Moguls were the fastest-growing market segment: exhibition.

The model of dual isolating mechanisms is shown in Figure 5.3. The model starts with the institutional resource containing the initial conditions and the building blocks of each network [1]. The Trust and the Moguls applied quite different templates into the same context, thereby creating

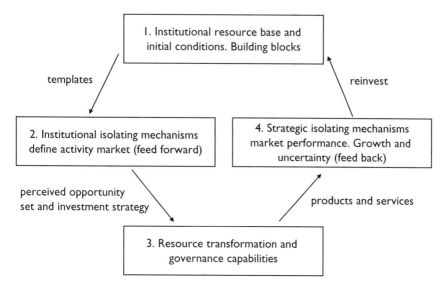

FIGURE 5.3 Strategic and institutional isolating mechanisms: co-evolution (from: Candace Jones 2000)

isolating mechanisms [2]. Their different perceived opportunity sets and investment strategy shaped the transformation of resources into capabilities and structure [3]. They produced different services and products (e.g. types of film). Therefore market performance [4] in the face of growth and uncertainty affected their opportunities for re-investment (feed forward) into the institutional resource [1].

Global corporations

Giddens (1990) defines globalization as the intensification of worldwide linkages, penetrations and extractions that selectively interweave and link distinct localities through expert systems and symbols (e.g. money) whose structural features possess unintended outcomes and are significant, relatively autonomous econo-political circuits. Explaining the distinctive features of the contemporary situation requires the Braudellian perspective of centuries rather than decades. Arrighi (2000) explains globalization through three questions:

- What is new in the present wave of globalization?
- If there are novel features, can they be theorized as elements within a social trajectory of evolutionary pattern?
- Is there a departure from past patterns of recurrence and evolution?

Arrighi contends that there is a novel pattern and an apparent relocation of the epicentre of the global economy.

This process of de-centring modernity has unfolded rapidly in the USA, yet the USA was the arena for the constitution of modernity. The transformation in the Cold War introduced a degree of disorientation for the USA as the chief agent of modernity. Despite Fukuyama's (1992) claims about the 'end of history', his work acknowledges the inability of the state to shape the aspirations of individuals. The counter-project had mainly been labelled post-modernity. Theorizing the transition to post-modernity requires developing concepts that explain the *reflexive construction of reality*. Reflexivity is attained by uncovering the structures and mechanisms whose transformation inserts novel alternative futures with different relationships from those previously experienced. To grasp the nature of the transition requires a grip on the trans-historical and cross-cultural potential of theory coupled with a phenomenologically rich analytic description involving novel concepts (Albrow 1996).

The debate of whether the transition in capitalism is a rupture or a heightened continuity has focused upon the notion of globalization. Being global means inhabiting a different spatial context and the relational configuration of 'forces' that shapes the zones of manoeuvre for corporate agency.

Invoking post-modernity and the global age means jettisoning three centuries of assumptions about the direction of history. It is no longer useful to deploy conceptual schemes anchored in the notion of a nation state because free-floating sociality is released in world society with global financial cities: New York (equity markets), Tokyo (banking) and London (foreign exchange). No single centre controls.

Globalization involved the creation of transnational agencies, practices and transnational professional associations and a capitalist class (see Sklair 2000). Transnational companies (TNCs) play a direct and indirect role in the circulation of all kinds of investment. Forty of the world's largest TNCs have over 50 per cent of their assets overseas and the average for the 100 largest TNCs is around 40 per cent. The developed countries dominate both through their high international connections and through their relations with the less developed nations. Ten nations provide more than 80 per cent of overseas direct investment. The major investment flows have been between the USA and Japan in the J–USA connection. The Japan connection is several times stronger than the USA–Europe connection.

Globality emerges as a new level of organization where there is no organizing agent – no single sovereign power – yet globalization is not disintegration. The global age creates new activities and introduces novel areas of indeterminacy that explore the limits of previous forms of expertise and provide potential arenas for new kinds of political expertise. The global age promotes the global managerial class in respect of those working within specific large multinational businesses (MNBs) and of those in the core interstices of inter-organizational networks, as well as those working in international consultancies. These constitute a global professional class. The global professionals extend their general claim to exercise judgement in indeterminate situations into three new situations:

- The fact of international organization
- The handling of indeterminacy arising from the global spread of information and communication technologies
- The roles of corporate management.

The latter has revealed systemic problems in software system design, providing an opening for firms such as SAP. Consequently, whereas previously the government in the nation state provided the functional definition of professional work, there is the emergence of a worldwide arena for professional codes governing client–supplier relationships. The new information and communication technologies dramatically extend the potential for longer economic chains through time–space networks that are impressive although also subject to risk. For example, supermarkets extend the food chain so that produce from one national context is flown around the globe on the assumption that it is free from damaging materials. The development of these long networks requires both organizational embedding and new disciplines as well the hardware and software. Thus, MNBs create a new managerial class that moves around the globe.

Governments in nation states find that whole areas of economic activity are shifted to the domain controlled by MNBs. National boundaries are conditional rather than absolute. Meanwhile MNBs are increasingly exploring their role in shaping national culture. For the new state, national culture is commodified as a resource. There are many examples. The notion of 'Cool Britannia' expresses the attempt, briefly, by the Blair government in Britain to theorize culture on multiple levels, ranging from new Millennium events to the aestheticization of every action. These developments signal a rupture rather than a continuity of the previous world and are manifest in the capacity of capital to commodify and colonize culture and even intimacy.

Globalization is an increasingly autonomous, structured, cultural-ideological and economic long-term process of mutual interaction and encounters among different power networks (Roudometof and Robertson 2001). Global frames of reference emphasize the multi-faceted, multi-directional and complex power balances between nations and firms involving a multi-layered flow of capital, technologies, goods, cultural forms and scapes (as Appadurai 1986a). Every aspect of reality is intertwined and located within global processes. By the late twentieth century the multiple world systems studied by Braudel (1982) became interwoven and possess similar processes. Appadurai suggests that these interdependencies invoke cultural contests. There is sameness and difference. These are cross-cultural processes encountering differences of traditions in an infinitely varied contest.

The spatial and temporal dynamics of globalization have challenged social scientists to construct process concepts, yet the depiction of flows is highly problematic. Notions of nomads, co-mingling both selective elimination and selective emulation, challenge the homogenization thesis. The concept of the commodity chain (Gereffi (1994) and Korzeniewicz 1994) has been widely used to express the interdependent linkages from primary producers to end markets. The commodity chain also reveals the asymmetrical economic exchanges between nodes in the assemblage. Hence the global community chain is re-conceptualized as the knowledge chain (see Chapter 10).

Globalization refers to the growing number and variety of corporations that organize their profit-making activities across state boundaries in a decentralized system co-ordinated through the market and through financial capitalism. There is heightened competition between geographically distinct contexts, even between those within the same national boundary. Thus there is an increased geographical range of consequent social interactions and increasing proportions of those interactions are across international boundaries.

Arrighi (2000) argues that in previous waves of globalization there was an historical tendency for the financial surpluses to be located within those military power centres able to exercise hegemony. However, in the contemporary situation, the military power lies within the USA while considerable economic power lies outside the USA within, for example, South East Asia. Possibly there is a new form of leadership. There seems to be a fundamental reconfiguration of the relationship. Contrary to theories of imperialism, the wars have been controlled. Moreover, the massive three-fold growth of American international corporations since 1980 has been accompanied by their retaining surpluses and not enabling their acquisition by the state. This is a *novelty* in state-to-capital relations. By their move offshore the multinational corporations expressed no confidence in the capacities of the USA. The Reagan administration did acquire those surpluses to reduce national debt but actually increased debt. The radical right engaged in the role of deflating the social power of whole sections of the workforce in the First World when faced by powerful and potentially volatile capital. There was an offensive against the social rights of employees. Twentieth-century USA was quite unlike nineteenth-century Britain. The latter possessed and used its Indian Empire to cover the deficits on the balance of payments while policing the world. The USA does not possess an obvious equivalent to the Indian Empire. So, the novelty seems to be the fission rather than the fusion of military power and global financial power. There seems to be a relative balance between multiple centres in the world system. If so, then the novelty appears more similar to the world in the thirteenth century.

The central players are the large and growing number of multinational businesses because they are the sites within which and between which new activities are being connected through forms of network organization that differ shparly from previous forms of organizing. The share and form of world trade occupied by the multinational business has been transformed since the 1970s. MNBs occupy three-quarters of world trade, of which between 40 and 50 per cent is internal to the MNBs. Most MNBs achieve turnovers exceeding one billion dollars, of which more than one-third is gained from overseas markets. This figure is very much higher for MNBs from the small nations (e.g. Sweden, Denmark and The Netherlands). Their international investment is widening and deepening, and this has considerable implications for governments of nations which face the dilemma of attracting foreign MNBs while 'protecting' domestic MNBs.

Global markets involve a concentration around a small number of key MNBs in each market area and their unstable oligopolies. MNBs provide products and services that are systematically generic (e.g. similar platforms for different BMW saloon cars and SAP software) and also adapted to local conditions. Because these products and services are systemic their delivery requires the connecting together of arrays of organizations. That tends to create powerful core technologies which in turn act as sources of convergence in innovative directions (e.g. VHS) while creating a demand for 'best standard benchmarking' within the co-operative alliance. There are, therefore, considerable problems for new entrants and for older MNBs entering new market domains because globalization dramatically increases the costs of entry to different markets. Thus, small European firms intending to globalize (e.g. Tetrapak from Sweden) face the problem in the USA of priming their rivals.

MNBs are highly sophisticated in the use of diverse forms of expertise and in their potential capacity to create new arenas of activity that cannot be regulated by existing expertise and require the production of new expertise. For example, strategic alliances involve legally consequential relationships that depend upon detailed and scopic auditing of resources and performance. So, new roles have been created for alliances of accounting and legal firms.

Firms: zones of proximal development and manoeuvre

Firms occupy a given ecology. Global firms gain an influence over the ecology of co-operation and competition.

Ackroyd (1995) explores the interface between the societal context and the zones of proximal development for firms in particular societies, as shown in Figure 5.4. The diagrammatic flow of casuality assumes a pre-existing structural and cultural repertoire similar to that proposed earlier in Figure 5.1. Firms are conceptualized as mediators with differential access to rules and resources. Within the firm there are strategic alliances with varying degrees of effective agency and capacity to manipulate organizational resources. They are embedded in systems of political and social relations that shape the extent to which the firm is capable of reflexivity and possesses the capacity to manoeuvre. The constitution of the firm is shown on the right-hand side. The firm is situated in the *long durée* of its own, of the sector and of the national social capital and is therefore shown as feeding back (left-hand side) into the pre-existing structural and cultural repertoire.

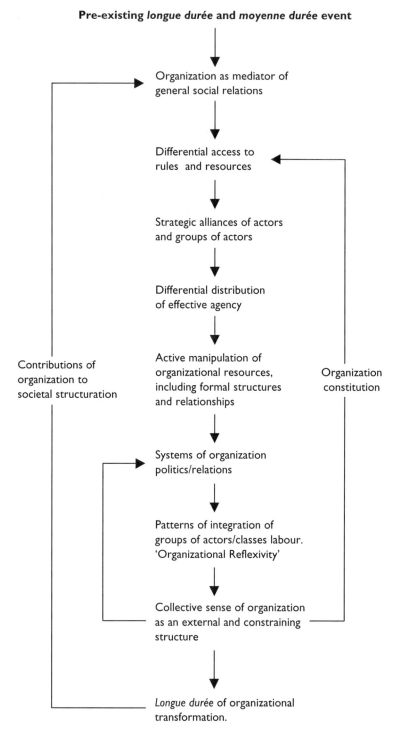

Pre-existing *longue durée* and *moyenne durée* event

Organization as mediator of general social relations

Differential access to rules and resources

Strategic alliances of actors and groups of actors

Differential distribution of effective agency

Active manipulation of organizational resources, including formal structures and relationships

Systems of organization politics/relations

Patterns of integration of groups of actors/classes labour. 'Organizational Reflexivity'

Collective sense of organization as an external and constraining structure

Longue durée of organizational transformation.

Contributions of organization to societal structuration

Organization constitution

FIGURE 5.4   Firms' zones of proximal development and manoeuvre (modified from: S. Ackroyd and S. Fleetwood 2000)

This framework highlights both the zone of proximal development for the firm and draws attention to how firms vary in their capacity to reformulate pre-existing zones of manoeuvre. The British case is one in which the pharmaceutical sector has performed well in the global economy over the past two decades. Also, the British hospital sector favoured the use of drugs more than of technology. Moreover, the forms of regulation in the health sector enabled research-based pharmaceutical firms such as Glaxo to prosper.

This chapter seeks to amplify the neglected yet central role of the market and of consumption in the new political economy. The chapter re-interprets the diffusion of innovations perspective addressed in Chapter 7.

The market as a quintessentially modern institution is now positioned in the new political economy having travelled through a post-modern passage. According to Baudrillard (1975), the process and meaning of consumption is the axis of culture. So, the market has a strong claim to be *the* arena of meaningful social action. The expanding hegemony of the market shapes consumption patterns so that 'theatres of consumption' may be understood through the metaphor of 'consuming people' (Firat and Dholakia 1998). Jameson's (1991) exposure of the post-modern complexion to the contemporary era acknowledges that marketing and consuming are central signifying processes. Marketing is more than a business activity. The consumption patterns of post-modernism are culturally diverse and are mediated by the market and the performative. In the theatres of consumption, the central struggle is between marketers and consumers. Marketers are concerned with market value and market share, while the consumer is concerned with expressions of meaning, self and culture. Yet we know too little about how consumption patterns evolve or about the role of central corporations and media players such as Hollywood in shaping consumption. Although the American economy has been defined as exceptional, we know little about its role in generating key elements of evolutionary variety and in securing their diffusion.

The discipline of marketing has freely used the notion of 'consumer systems' to situate the consumer in a flow of consumption opportunities. The notion of consumer systems is in a tournament with alternative conceptions because within marketing there are strong attempts to re-theorize the consumer system (e.g. Ambler and Styles 2000). The leading edge outside marketing is illustrated by the adoption of anthropological approaches to consumption from Appadurai (1986a,b) and Miller (1997a,b). Also, the sociological approach of Bourdieu has challenged the assumption that consumers purchase brands (Holt 1997). From Bourdieu it is argued that it is the way that the purchased product functions in use that is significant for the consumer. According to this perspective, consumers construct knowledge about their use of products and services and this knowledge is deployed in the consumption situation. Moreover, consumers are increasingly aware of the constraints, couplings and opportunities that arise from their own space–time biographies and marketers constructing 'consumer systems'.

Storper and Salais (1997) explore the production–market linkages through the theme of conventions of co-ordination and suggest four major types that are likely to be found in the new political economy.

Market society in the cultural turn

The key conceptual building blocks – individual, society and rationality – of the management and social sciences have been transformed by their encounter with theorizing the market society. The category of market society now shapes our imagination of social order and of modernity. In the past two decades the post-modern movement on the one hand and the new practices associated with marketing on the other have contributed a more nuanced account of the social structuring of the market (Clark 2000). Now it is necessary to understand the contending conversations. This section examines the mix of social mechanisms, including the market, without presuming the dominance of any single one.

The new debate centres on the 'cultural turn' and is a critique of the orthodox.[1] Orthodox debate centres on how social processes are co-ordinated. What role does a price mechanism play? What is the relationship between state planning and markets? We will examine the orthodox debate through two rather different accounts: the first consists of Braudel's historical account of the pre-modern market place and the modern capitalist market; the second, through the position that Weber (1978, 1982) and Parsons (1951) accorded to the economy.

First, Braudel (1982) traces the transition from concrete market places to the abstract modern markets where flows of money are largely hidden. Market places are still found yet seem to have disappeared in advanced economies. Market places possess a very public social centrality, involving the meeting of buyers and sellers within a cultural framework of rules about transactions (Braudel 1982). They are in a complex relationship to local and state governance (e.g. taxes) while providing arenas of liminality. They can be dangerous. In contrast, modern markets are located in networks that operate behind, above and between market places. The network and connections are largely hidden from the everyday gaze but are constantly under the gaze of the global service class: the eye and gaze of power (Chapter 3). The networks consist of global flows (e.g. trade, finance and information) with a vertical integration of trading. The abstract financial markets were embodied in trading rooms inscribed with distinct architectures and modes of sociality. Historically, in the case of England, the expansion of

---

1   This section draws from the excellent reviews by Don Slater (London School of Economics). See especially Slater and Tonkiss (2001).

London markets drew in most of the United Kingdom and related areas of foreign trade.

What happens to the market place? From the post-modern perspective it persists in the simulated, themed shopping mall and the transformation of the supermarket by the introduction of counters in the form of market-like stalls. The concrete modern shopping spaces are choreographed by their owners to replicate market places. Some spaces are given oriental themes. Overall there is the chaotic-exotic display similar to that employed by nineteenth-century American department stores (Butsch 2000). Behind them are the markets for private finance and capital.

Second, the most commonly accepted model for markets is taken from economics where markets are integrated systems of exchange where actors can pursue competing interests that are reconciled through the price mechanism. Prices co-ordinate markets and the presumption of the model is that behaviour is calculative. This model takes different forms, for example, between Adam Smith and Marx who promoted a monolithic model of the market society. That model influenced Weber and through Weber into Parsons and thence into organization theory.

The founding sociologists occupy a major position in this economics-dominated framework. There is a strong line from Weber to Parsons. Weber theorized the market agent as an ideal-type of formal rationality and proposed instrumental rationality – efficiency, calculability, quantification, control and prediction – as the most sensible mode of operating. The theorizing of the economic, and therefore of the market, by Parsons has been influential in social science. The economy, as per Weber, is embedded in the social order as on the universal functional problems that all social systems encounter in their experience of social order. The economic is intrinsically social, cultural, political and normative, yet also the most rational system. Therefore the parameters of the economic shape the other sub-systems.

What then is a market society? Three features characterize them:

- A high technical division of labour coupled with social integration
- Commodification for profit in the capitalist framework of calculation across the whole reproductive cycle. Commodification depends on markets, yet is also part of the 'social life of things'
- Monetarization coupled with calculation/quantification and depersonalization. A money price enables the quantification of alternative courses of action with local definitions of rationality.

Market society became a new social actor. Therefore economic order is always embedded in society, but how does the cultural turn re-theorize the role of the economic? We now examine the cultural turn.

The cultural turn

The more social approaches contend that social networks, social structures and meanings construct economic institutions, so markets are social institutions with affective, social rule-governed, customary, conflictual and socially rational influences. Economic action is irreducibly social and economic institutions are social forms. Exchange has to be theorized as social rather than as narrowly economic and individualistic/systemic. Consequently, the disembedded model of Parsons does not explain the transformation from the market place to the abstract market and that economistic fallacy must be replaced by considering market exchange in the context of reciprocity and redistribution. Even contracts exist only because of generalized exchange and the circulation of 'symbolic' credit. The market is both a social artifact and also a societal mechanism.

Non-economic factors such as social ties and cultural norms shape the processes of material allocation. Moreover, even non-market exchanges might involve tight degrees of strategic calculation by actors and political groups. So, 'embeddedness' has to be unpacked to highlight the cultural norms, the institutional structures and the political frameworks and internalized rules.

Modern societies draw attention to how economic activity might be seen as embedded in the networks and complex transformations of social life over the past three decades. There has been a speeding-up and a stretching-out of exchange processes over greater distances (see Chapter 3). Boston supermarkets receive their perishable products from around the world. The stretching of networks across great distances means that local embeddedness may mean the demise of trade union influence, coupled with the survival of the parent-teachers association. Social capital may be very selectively retained (see Fukuyama 1995). Therefore it is important not to obscure the market in some organic notions but instead to address the interface between economics and sociology. The institution of markets involves the role of laws and policy in shaping markets; for example, the quite different conceptions of social market for local medical care in Britain between the Thatcherist notion of fund-holding and Blair's New Labour. Also, legal strategies shape the use of space. German rules on retail trading may inhibit foreign entrants such as Wal-Mart and Virgin.

What is at stake is a struggle between cultural practitioners and their audiences to legitimate their own social, cultural and economic capital in the social market place. Cultural competition to legitimate aesthetic capital in the market society can be theorized in terms of the strategic and 'rational' use of resources. False universalism and false particularity characterize contemporary culture because the cultural industries subsume various cultural products. The cultural industries produce their products through rationalized processes (see Benjamin (1999)). Thus the apparent diversity of cultural choices reduces to similar experiences. The formula pursued by the cultural industry is seemingly to customize while extensively relying on multiple common platforms and modules. The BMW car is a triumph of standardized variety. Shopping is a major site and a revealing focus for produced cultural experiences (Campbell 1996). Shopping can be depicted as a consumable spectacle of objects and sociality where the consumer encounters the commodity. In the recent theorizing

– drawing on Walter Benjamin – shopping is the space/place of many diverse pleasures in addition to the act of buying: gazing, fantasy and desire. All these are promoted by the brandscapes. Within capitalist culture hedonist shopping might be construed as legitimating utopian moments. And this has particular implications for women. Women can traverse these spaces. Shopping can be an empowering resource.

The cultural turn profoundly reformulated a whole cluster of issues including social order and individual identity, as well as the claim that economics is a cultural discourse arising in a particular problem of governmentality. Slater and Tonkiss (2001) contend that the cultural turn recognizes and extends the standard culturalist critique of the economics theorization of the market by the post-moderns and by critics such as Williams (1985). The cultural turn treats economics as a cultural item.

This examines the central role of the symbolic systems and an informational process of cultural processes and institutions constitutes the economy and the material things that flow through the economy. So the economic is contextualized in the cultural. In this format it is the consumption of cultural materials that creates the circulation of the flows in the 'economy'. The cultural turn points to the increased centrality of marketing in the management of economic institutions and processes through the aestheticization of everyday life by, for example, marketing, design and life-style imagery. It is cultural logics and cultural goods that increasingly comprise market economies. The cultural and the economic have imploded and the economic flows are of signs, information and cultural dynamics. Cultural experts occupy a key role in conceptualizing, knowledge, design and marketing. The cultural turn places the market in the centre of social life. Marketing creates the cultural specification and significance of goods. Marketing is increasingly grounded in the development of just those kinds of knowledge that commodify and colonize social experience and meaning. The market plays a role in the formation of identity and construction of meaningful social life through providing symbolic resources that develop difference and distinction. This involves a new way of labouring for capital because many of the skills required are close to the self. One has to present oneself. Many television shows exemplify this feature through the central location of non-celebrities (e.g. *Blind Date*, *Big Brother*). There is a tight connection between cultural production and consumption. So markets are flows of signs: brandscapes. There are many non-material goods and the non-material component of material goods has increased through commodity aesthetics and sign values. The advertising media and retail spectacle symbolically mediates goods in economic processes. The signs are cognitive and aesthetic, and an image becomes embodied in the material object. All stages of a commodity's life cycle – design, production, circulation and consumption – are aestheticized. Consumers are implicated in the spectacles of display and consume these. The role of meaning and symbols in commodity circulation has been heightened in terms of identity and social place. The cognitive signs drawn from cybernetics shape our thinking about flows.

In the cultural turn the enculturation of the market can be theorized by the post-moderns (e.g. Baudrillard) as an abstraction that feeds forward to engulf and transform the spaces and places of market capitalism. So, the former attention to the 'use-value' of products is displaced by their 'sign-value' which is derived from its position within codes of meaning and semiotic processes. The creation of sign-value becomes a specialist area of knowledge: commodity aesthetics. Aestheticization constitutes the market through an abstraction (cf. Marx) in which the distinction between reality and image is effaced. Capitalist post-modernity generates an aesthetic coating of images and social order is achieved through relations of meaning because the sign affects the social bonds. The logic here is that the market would disappear because there are no realities for the market to represent. Instead, the shopping malls simulate the market place, leading consumers to believe that they live in a market society. Consumer culture represents the manipulation of the consumer because identities are negotiated through consumption and commodities. Consumer culture as a feature of capitalism involves an enormous growth in specialists dealing with 'signs', advertising, media, leisure, moulders of meanings and life-style centres. These meanings and signs are in constant revision, thereby impacting the notion of the reproduction of everyday life.

For cultural analysts the notions of 'flows' and of 'networks' subsume various social processes and inspire renewed interest in the spatio-temporal. Appadurai (1986a and b) conceptualizes these flows and networks as imagined worlds with five scapes: mediascapes, technoscapes, financescapes, ideoscapes and ethnoscapes. The scapes constitute different social landscapes. They are internally driven, interconnected and yet disjunctive. The scapes go far beyond Braudel's account of the market place and of the abstract market. Thus international trade constitutes markets as one aspect within multiple flows. The spatio-temporal extension – time and space distanciation – involves growing mobility across borders to create continuous hybridization rather than homogenization. Hence the earlier assumption of capitalism that there would be a global market is replaced by the metaphor of creolization.

The cultural turn redefines the role of culture in economics so markets are social events with social actors. However, the culture of economics provides a powerful discourse and has real consequences (e.g. Thatcherist markets). The crucial element is the enhanced role of the media in shaping images and scapes (Dodd 1995). Greenspan's 'rational exuberance' is based on a media-transmitted image of market-driven enrichment through share buying. Therefore we need to redefine economics to include the knowledges and practices found within market institutions: marketing, advertising, management. Slater and Tonkiss, and Callon observe that economics operates as a structure that constitutes the spaces in which market actors are produced and framed. Economics is a discourse and social technology that constitutes things – including markets – by 'making itself true' in a self-fulfilling way. Economics might therefore be theorized as a form of governmentality in the control society.

Commodities and tournaments of value: Appadurai

Appadurai (1986b) asks under what conditions do economic objects (e.g. flowers and computers) circulate in different regimes of value in time and space? To explore this question he develops a perspective on the circulation of commodities. This starts from the claim that economic exchange creates value and the value is embedded in the commodities that are exchanged. The link between the exchange and value is through politics and the social life of commodities. Therefore *commodities have a social life*. The demand for an object endows the object with value and so exchange sets the parameters of utility and scarcity. Exchange is therefore the source of value.

Economic activity is cultured in commodity situations. To theorize this Appadurai incorporates the parallels drawn by Bourdieu between the temporal dynamics of gift exchange and the strategic and calculative elements conventionally associated with economic activities. There is calculation in both activities because practices always conform to interested calculation even when they appear to be disinterested. Thus economic activity is cultured in commodity situations. Appadurai (1986b: 13) defines a commodity situation in the social life of any 'thing' as the situation in which exchangeability (past, present or future) for some other thing is socially relevant. The commodity situation can be disaggregated into the commodity phase, commodity candidacy and commodity context:

- The commodity phase of the social life of any thing is when 'things' are typically moving into and out of the commodity state. This movement may be slow, as in traditional societies, or fast, as in contemporary capitalist society. The movement can be terminal or reversible and normative or deviant. There is a biography of things.
- Commodity candidacy refers to the standards and criteria that define exchangeability in a particular geo-historical context, for example, comparing the regional policy in Southern Italy with Silicon Valley. The criteria are symbolic, classificatory and moral.
- Commodity context refers to the variety of social arenas within and between the cultural units of analysis (e.g. British Motor Sport) that link the candidacy of a thing to the commodity phase of its career (e.g. the 2001 Ferrari). Arenas are diverse. In some societies marriage transactions constitute arenas in which women are intensely and 'appropriately' regarded as exchange values. Auctions accentuate the commodity dimension.

Commoditization is therefore ubiquitous and involves a complex intersection of temporal, cultural and social flows. Appadurai introduces further distinctions between commodities in terms of their:

- Destination
- Transformation

- Diversions into and out of commodity states
- Ex-commodities
- Interpretative flexibility in usage (low/high)
- Necessities and luxuries.

The flow of commodities is always a shifting compromise between paths that are socially regulated and diversions that are competitively inspired. There is therefore both regulated improvisation and also the potential for transformation.

The politics consists of tournaments and contests of value which:

- Signify the relations of privilege and social control
- Contain a tension between the existing framework of bargaining and price and the current transaction
- Reveal the tendency of commodities to breach the existing framework.

Therefore contests of value can loosen existing rules. Examples can include the new dress codes for accountants and consultants and the aim of Ted Baker to stop men wearing ties. Tournaments of value are complex cultural and periodic events that are separate from the routines of economic activity and yet have subsequent consequences for those routines. The tournaments tend to be played through by those in power and involve status contests. These tournaments can be between alternative ways of co-ordinating the firm through the software of information technology. The tournaments are about more than the status, rank, fame and reputation of actors because they engage with the central tokens and signs of value in a particular society. The tournaments involve strategic skills and some actors will attempt to create diversions from the established paths for flows of 'things'. Tournaments are complex processes for *re-calibrating the biographies of things*. They take varied forms and involve the politics of display.

Appadurai's political economy of consumption differs significantly from much of the existing consumer research. Consumption and demand are located in the overall political economy. Demand is conceptualized as being primarily shaped by the variety of social pricks and systems of classification in a particular society rather than being the 'mysterious emanation of human needs' (1986b: 29) responding mechanically to either social manipulation or some universalistic desires. Demand is the economic expression of the political logic of consumption and consumption is social and relational rather than private, atomic or passive. Thus the universal human needs and utility theories are rejected by Baudrillard and replaced by social mechanisms within a relational theory. This can be illustrated with reference to fashion.

Fashion suggests rapid turnover, the illusion of total access, coupled with a marketized assumption of a democracy of consumers (Appadurai 1986b: 32). There are mechanisms through which certain establishments control fashion and good taste and limit social variability. Thus modern consumers are victims of the velocity of fashion to pursue certain commodities. Moreover, even when consumers possess discursive penetration about the shaping process (e.g. through media humour, irony and pastiche), there are effects. In capitalist societies our consumption patterns are regulated by a high turnover in the criteria of appropriateness. Consumer demand is a socially regulated and socially generated impulse rather than an individual whim.

Commodities such as CDs and cut flowers are involved in complex flows across different cultural boundaries and so there are always potentials for discrepancies in local knowledge about commodities. What are the peculiarities of knowledge that accompany action at a distance and how does the tension between knowledge, ignorance and resistance critically determine the flow of commodities?

Commodities are complex social forms and distributions of knowledge. Three forms of knowledge – technical, social and aesthetic – can be distinguished in each of three different, connected arenas:

- Production (e.g. making Formula 1 racing cars)
- Consumption (e.g. watching Formula 1 spectacles)
- Intermediaries.

It is too simplistic to consider technical knowledge as being only within the production arena or aesthetic knowledge as being located only within the consumption arena. All three knowledges are found in varying forms in each of the arenas: there is mutual and dialectical interaction. By constructing the biography of any commodity (e.g. racing car, carrots, napkins) we can examine the dynamics of knowledge distribution at the different stages of design and production.

Appadurai considers that culturally standardized recipes may dominate the production locus, but the study of the production of Formula 1 racing cars suggests some variability. For example, the knowledge required to design an F-1 Grand Prix course has to accommodate both the semiotic features that make the race a spectacle for the global media and also the local spectators. What is of note is that in the 'new cultural logics of capitalism' (Jameson 1991) a key area is knowledge about the market, the consumer and the destination of the product. The post-modern movement has highlighted the issue of specialized knowledge and authenticity. Capitalism as a techno-economic design is a complex cultural system whose specific western history involves geographical difference within the USA and between the USA and other societies. In the USA the cultural design of capitalism (e.g. museums of the future) has been vigorously explored because commodities and meanings play a considerable role.

Appadurai notes how the template of futures markets established in Chicago in the 1880s is an institutional arena where risks about the future can be hedged. These are speculative tournaments seemingly divorced from the everyday routines or at least permitting some insulation for certain periods. Their game-like ethos merits further attention.

In complex capitalist societies the relationship between knowledge and commodities involves both segmented, distributed knowledges and knowledge about commodities which itself is being commodified at an increasing pace. The commodification of knowledge involves different roles for expertise, credentialism and high-brow aestheticism. There is increasing similarity between commodification of goods and services. The role of advertising illustrates the relationship between knowledge and demand. It seems unlikely that advertising is as simplistically effective as some neo-Marxist analysts have supposed. The images can be analysed as a species of 'capitalist realism' which is a cultural representation of capitalist life-style in general (Schudson 1984). Contemporary modes of advertising shared a common strategy in the 1980s of making ordinary, mass-produced, cheap commodities seem desirable and reachable. The images of sociality aim to suggest that the commodity be bought as an afterthought.

To conclude, as commodities travel greater distances, then knowledge about the commodity becomes differentiated, partial and possibly contradictory. There are shifting arrays of commodity paths involving both the politics of diversion and enclavement. Where smaller systems interact with larger ones, the larger system acts as a turnstile influencing the flows.

Politics and contests implicated in power link value and exchange in the political life of commodities. The political process signifies and constitutes relations of privilege yet commodities also breach those relationships. It is possible that at the apex of societies the tournaments decide the fate of new commodity paths.

Consumption as the vanguard of history: Miller

It is likely to be the bourgeois female . . . who will become the progressive force in the first stage of the new millennium.

(Miller 1998: 49)

The social ethnographer, Daniel Miller, examines the role of consumption in the new cultural logics of capitalism. He attacks two claims. First, that the ground of the contemporary world was established in the earlier production area. Miller cautiously argues for the influence of retailing. Second, he confronts the claims of the marketers and their radical-left critics, who obscure the role of consumption as a political activity. He contends that demand-led capitalism can help the capitalists search for profits. It is the aggregate decisions of the First World consumers, especially housewives buying in supermarkets, that are relayed on a daily basis as the nature of demand. The consumer 'votes' on a daily basis, but there is no democratic deficit for the Third World suppliers. These consumers can largely ignore seasonality and can experience a real drop in the prices of food relative to the past. The retailers can use information technology to couple supply and demand and to customize their products.

It is the housewife who epitomizes the contradictions of contemporary power. Although she commands slight respect, her skills in thrift and comparative purchasing have considerable implications for the domestic quality of life. Miller grounds Foucault in its materialist base by an archaeological exposure of the limits of economic theorizing of the market and by examining four myths of consumption:

- That mass consumption causes global homogenization or heterogenization
- That mass consumption is opposed to sociality
- That mass consumption is opposed to authenticity
- That mass consumption creates a particular kind of social being.

So, what is consumption? Miller contends that consumption is the attempt to extract humanity and to negate the generality and alienatory scale of the institutions and world they inhabit (1998: 31). Modern consumption is about identity and the housewife occupies a central role in the construction of identity. Therefore consumption is politicized even though – aside from the now departed *Marxism Today* – the major political groups failed to grasp this feature until the coming of Bill Clinton and Tony Blair in the 1990s. The rhetoric of the consumer choice is located in the local extraction of value: 'it is likely to be the bourgeois female . . . who will become the progressive force in the first stage of the new millennium' (ibid.: 49).

Miller rejects Marx's grand narrative and its homogenizing assumptions and contends that the dialectics were/are always grounded in the historical conjuncture. Therefore the narrative would alter with the emergence of new conditions such as the consumer society. There should be an unfolding analysis that shifts as each future is appropriated by theory. Thus, he concludes that Marx's utopian scenario was a rhetorical gesture (see Hodgson 1999). So, contemporary research should be more analytical, more intrusive and more observational. However, as Miller notes, there is very little research on consumption. We need to move beyond the duality of symbolic versus utilitarian found in consumer research into the link between emerging differentiation and new forms of consumption.

Theatres of consumption

What does 'theatres of consumption' mean? Theatres of consumption can be illustrated from Butsch's (2000) account of the making of American audiences by the transformation of leisure into consumption. In America, leisure activities have been commodified and commercialized into a purchased commodity such as television sets, tickets and equipment (e.g. baseball). Butsch contends that process has been continuous and uneven for two centuries. In the nineteenth century the local entrepreneurs supplying class-based tastes were displaced by national oligopolies and mass media (e.g. the role of Spalding in tennis). He contends that *suppliers became highly influential – even shaping the rules of games* to make them more desirable to consumers. So, have consumers become tamed by the hegemony of commercial leisure? Certainly media research was promoted to provide capital with the capacity to persuade consumers to purchase goods and services. Butsch contends that there are issues of domination because everyday social practices need to be understood in terms of power and a non-deterministic, historically dynamic concept of domination such as hegemony. Hegemony entails class domination through the participation of subordinate classes. The hold of hegemony varies and is never total. If elites can define the structure of leisure practices, then they have hegemonic influence (e.g. passive sport watching). So, if people produce their own leisure, then hegemony is low and vice versa.

Leisure activities require structure and meaning. Commercialization of leisure practices involves their entrainment to the economy. This has unfolded in America in three significant phases: to 1830, to 1880, and since 1880. Each period has a distinct form of hegemony. First, traditional authority was highly paternalistic through ecology of spaces. Compare the spaces for the Pittsburgh elite who could attend four quality cricket pitches with the spaces for non-elites. Second, leisure becomes an arena of class struggle with entrepreneurial capitalism in small town America (e.g. theatres, sports). Because of the market relationship (i.e. paid for consumption), the non-elites possessed some influence, but their capacity to structure activities was limited, especially in sports such as American Football that emerged at the end of this period. Theatre audiences were continually transformed from spectators into consumers. Merchandisers in New York progressively increased their power base in production and distribution. This would later form the institutional block for entering the movie industry (see Jones 2000). In this period the position of 'sporting goods' became more and more central. As goods – bats, balls, spikes and hats – they were physical expressions of ideology. These new providers determined much of the leisure revolution. The marketed goods built up a structure of authority like 'coral islands' (Hardy 1982: 91). Third, after 1880 leisure becomes consumption: 'leisuring for capitalism'. Major developments in the leisure industries unfolded after 1880 with breweries and, soon after, the movie industry. In all of these 'new marketing' strategies were introduced and these preceded or went hand in hand with oligopolistic concentration. This included the standardization of rules for the games and hence for their equipment supplier. The movies were an important vehicle for these developments. There was a gradual development of an appropriate audience behaviour that might be summarized as 'middle-class decorum'. There were some surprising contributors. The US Forest Service, for example, was a major agent in the commodification of outdoor recreation. Leisure practices became increasingly capital intensive because recording and amplifying technologies centralized the production of entertainment and replaced the itinerant 'armies of entertainers'. There was competition between types. Between 1926 and 1929 radio audiences doubled and piano sales fell by almost 60 per cent. Television in the 1950s was accompanied by falling cinema audiences. Television colonized leisure, with the average household watching more than 35 hours per week between 1955 and 1985.

Leisure practices demonstrate that the oligopolies wielded and imposed considerable influence – hegemonic power – but not total power. Even so, there is a semiotic materiality to the making of American audiences (Butsch 2000).

Catwalks, brandscapes and female identities

The cultural turn draws attention to the role of the media and its construction of brandscapes to seduce the consumer in the new political economy. Satellite television and the potentials of digitalization enable extensions of capitalism into brandscaping as a means of colonizing everyday life with revenue generating images and signs. One example of this is in selling clothes and accessories to the market of young women. This is illustrated in Figure 6.1 with a model drawn from the analysis of the issue of whether the media is responsible for changes in female identity. In the bottom layer young women are shown (from right to left) becoming consumers. Their theatre of consumption is a relational configuration (see Chapter 4), significantly yet not totally shaped by the actors in the upper half of the model:

- Commercial television receives revenues from advertisers
- Powerful corporations advertise on television and seek to have their brands located in television events (e.g. sports)
  - Develop new raw materials for fabrics
  - Research consumer life-styles
  - Look for themes that excite consumers
  - Develop and test prototypes
  - Develop brands with sign values
  - Support and take part in catwalks and use celebrities
- Powerful cultural entrepreneurs link different arenas such as diets, catwalks and life-styles to articulate theatres of consumption
- Powerful retailers construct shops as theatres of consumption.

The actions of these powerful players are, of course, regulated by the state in various ways (see Chapter 5) yet leave the corporations with immense potentials. It is those potentials that create the brandscape of images for young females and contribute to their sense of rewards and stress for attaining certain profiles. Young female identities are displayed on the street, in the workplace, clubbing and possibly by being shown – momentarily – on television.

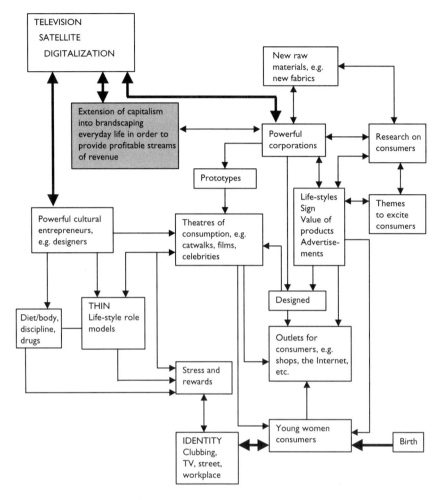

FIGURE 6.1   Theatres of consumption

The genre represented by this model can be applied to every other area of consumption, especially to track organizational innovation in public settings.

Conventions co-ordinatiing the political economy

Conventions as the cultural foundations of capitalism

Storper and Salais (1997) propose that in this phase of the new political economy trade and specialization are driven by the capacity of specializers on the supply side to learn, create and capitalize upon new knowledge. Therefore the design of the new outputs for consumption tends to be tightly articulated with their acceptance. Outputs from the supply side are continually redefined. Because the economic signals resistant to growth cannot be reduced to a calculable risk firms face a wide zone of uncertainty in their strategic choices. There is an extensive amount of localized information swirling around in the advanced economies, but the capacity of actors to notice and process this information is highly variable. Also, there are extensive untraded interdependencies between firms, sectors, technologies and actors and these narrow down the durable pathways for economic development. *Understanding these social pathways for future economic growth requires a thorough analysis of the cultural foundations of capitalism.*

The successful co-ordination of complex, heterogeneous ensembles involves interpretation, precedent and convention to discover and constitute coherent patterns collectively. *Conventions for co-ordination* represent a framework of foreseeable action for economic actors. To understand economic actors requires examining the conventions governing pragmatic situations and frameworks depicting collective action in economic activity. Because all economic activity involves coherence between the constraining structures and the conventions that enable action, there is a pragmatic problem of establishing and maintaining a common existence for the actors involved. Conventions define the common existence for all actors in a particular world and they face certain external requirements:

- Profitability is an essential measure of performance
- The size of the international market for traded outputs (e.g. consultancy)
- Identity refers to the taken-for-granted network of roles and this rests on deeply sedimented competencies and characteristics of groups
- Participation is how people conventionally act in particular spaces.

There are many possible ensembles, each grounded in a particular form of co-ordination between the diverse economic actors that have to be mobilized in order for a particular output to be sold. Co-ordination connects those who utilize the output (consumers) to those who supply the output and is mediated by innovation-design. To be efficient the co-ordination of the heterogeneous collection of actors must be aligned with the co-ordination of expectations. It is the conventions in the ensemble that make that ensemble possible or impossible.

Conventions are systems of expectation with respect to the competencies and behaviours of others. Conventions cover all the activities. Actors generate conventions around the enacted and pragmatic situations in which they find themselves. Within the ensemble collective actions are coupled and they collectively evolve in a (subsequently) recognizable trajectory that is specific, local and possibly durable. Learning co-evolves and because anticipations are so central to the ensembles there is a subtle experience of recurrence and temporality that is specific to particular places. So as situations recur, actors bring to bear their enacted repertoires.

Co-ordinating conventions: Four ensembles

There are four major forms of co-ordination between consumption and production. The notion of ensemble is deployed to characterize inter-dependencies within and between economic sectors. Storper and Salais (1997) construct four ensembles of input–output relationships from three dimensions:

- *Generic/customized outputs.* The market for outputs can vary from measurable uncertainty (or risk) to uncertainty where the risks cannot be calculated.
- *Standardized/specialized inputs.* The conventions and anticipations governing resource inputs vary. Specialists occupy a key position in co-ordination and the non-routine technology tends to be strongly equated with the specialist. A convention of standardization enables inter-changeable and reproducible resources. These two different forms of co-ordination affect supplier–consumer relationships.
- *Scale/scope economies.* There are different conventions for high volume and high variety.

These three dimensions are dichotomized to give four ensembles:

(1) The *Interpersonal Ensemble.* Specialized and customized products reflect the perceived desires of the consumers and the conventions, as in the fashion, design and craft-intensive work situations. The conventions rest on reputation, confidence and specificity of image. These conventions shape certain economic activities although they are probably less prevalent than in the pre-Fordist period.
(2) The *Market Ensemble.* Conventions for standardized and mass-customized products/services can be expressed in formalized, codified norms. Competition is based on price and the rapidity of supply. These are not universal conventions but are found in particular locales.
(3) The *Industrial Ensemble* contains generic standardized products whose co-ordination conventions are mediated by notions of efficiency. The possibilities for this world have receded.
(4) The *Intellectual Resources Ensemble.* The conventions for the co-ordination of creation have to cope with a diverse, multiplicity of production methods and types of output. Conventions are embodied in professional and scientific rules.

Each of these is distinguished by fundamentally different routines for enacting and absorbing uncertainty. The four ensembles are co-ordinated around different principles for connecting the market to its suppliers. The differences are in terms of the forms of uncertainty, the response to uncertainty, the basis of competition and the evaluation of quality.

To elaborate further, the conventions in the *Interpersonal Ensemble* of specialized, dedicated outputs with high uncertainty and the economies of variety locate the basis of competition in quality and the evaluation of quality is in terms of price. There is the extreme state of uncertainty where conventions indicate few points of reference to start the process between supplier and consumer. These design-based outputs include such diverse items as the barrister in a complex commercial dispute defending property rights. Great attention is focused upon meanings, experience and a common language.

In contrast, the conventions in the *Industrial Ensemble* of standardized, generic outputs are of predictable risk with economies of scale. Uncertainty is around the business cycle and fluctuations in demand and the response to those uncertainties will be in terms of forecasting the medium term. The basis of competition is price and the evaluation of quality is against generic industrial standards. The conventions about the constitution of outputs construct uncertainty as a predictable risk. Norms of quality are codified and the constitution of outputs is mediated by the material context of buildings, equipment and raw materials. In the *Market Ensemble* the conventions about outputs are of standardization, even when a run of production is dedicated to specific clients. The convention of standardization is located in the language. In the *Intellectual Resources Ensemble* the conventions guide the processes whereby the qualities of existing objects are altered. Action also proceeds in a context of uncertainty. Workers in this context survive longer when they develop knowledge that has applicability across a number of consumers. Their conventions involve local codification.

The commodity is a material outcome of a complex ensemble of social processes which involve the participants learning and aligning the multiple conventions that are specific to whichever of the four ensembles they are within. The four ensembles are mutually exclusive, but particular sets of actors and specific locations are finding that their future depends upon switching from one ensemble to another.

Firms probably choose the outputs that they consider will yield profits rather than having a theory of co-ordination. Perhaps the firm is not the best unit of analysis (Clark 1987, 2000). Firms participate in an international inter-organizational web of networks involving a division and distribution of knowledge. The firm soon acquires routines as a means of efficiency. For example, in the Marshallian market model, the specialized firms serve varied demands. These firms develop routines for searching for new clients and other routines to maintain flexibility.

The conventions for labour for each ensemble do vary (Storper and Salais 1997: Figure 3.4).

- In the Interpersonal Ensemble the firm deliberately groups individuals into networks but each person is responsible for the adjustment to unknown factors.
- In the Intellectual Resources Ensemble co-ordination is through small groups which have their own responsibility for developing knowledge.

- In the Market Ensemble co-ordination is of individuals who experience variations in the demand for their labour.
- In the Industrial Ensemble the firms possess internal labour markets and labour adjusts to the unknown through unemployment.

States and their repertoire of conventions

Storper and Salais (1997) examine the connections between their theoretical concept and the economies of sectors in France, Italy and the USA. In the French case, there were clear employment losses from the Industrial Ensemble, with firms moving towards the Market and the Interpersonal Ensembles. These shifts create problems of adjustment. Although there are multiple pathways from the Industrial Ensemble (Storper and Salais 1997: Figure 4.1), there is a constant threat of co-ordination failures because the surrounding conventions and institutions take shape to protect their actors from threat. Because Storper and Salais have attempted to see things from the standpoint of the actors themselves in the midst of situations (1997: 209), they emphasize that these work institutions emerge and are not created by the state.

The notion of situated states is introduced. In situated states, actors have the autonomy to develop whatever ensembles they find compatible with their frameworks of action. There is a *zone of organizational choice*. Autonomy is defined by reference to collective action (cf. individual autonomy). In this perspective, the state's role is to ensure that the frameworks of action and their assemblages of practices for co-ordination are treated with respect. There could be a variety of ensembles in the same state. However, societies may have a potential only for a limited variety (Clark 2000). Storper and Salais (1997) use the notion of situatedness to reject holistic and essentialist conceptions of society (cf. socialist utopias). Positive political economy is moving away from the state-versus-market distinction to the scrutiny of how different incentive and disincentive systems affect the actors in the specific institutional circumstances.

The effects of the state on the capacities of actors are crucial. The issue of choice hinges around the relationship between international competitiveness to *real worlds* and *situated states*. Therefore administrative rationality by the state is much more limited than current policy theories allow. They are confined to *an idealist notion of the institutional architecture* and omit attention to the processes *of co-ordination* and the frameworks of action. In effect, they use institutions as vehicles of policy to replace culture and that separation between institutions and culture elides the features of action and co-ordination. But it is practically impossible simplistically to adopt exogenously determined best practices. Conventions and frameworks are frequently problematic when existing routines do not match the tests of linking the external with the internal. This argument has immense consequences for the use of regional policy in Europe.

Storper and Salais reformulate the problem of the state's role in the light of their analysis of the future for Italy, France and the USA. In Italy there is a problem of combining the Interpersonal Co-ordination with scale. Italy is regarded as setting the standard for a situated form of industrial policy adapted to Marshallian markets (Marshall 1961). However, the design-led industries face challenges with specialized dedicated outputs and the necessity to keep deepening and refining the asymmetric knowledge that permits innovation. There is an exhaustion of traditional domestic entrepreneurial resources. Attempts to cope reveal co-ordination problems yet have not brought the end of decentralized production in the form of vertical integration. Although proximity operates, there are collective problems. The French face the problem of joining the Market and Intellectual Resource Ensembles via Interpersonal networks. In the USA the Intellectual Resources and the Industrial Ensembles are mobilized within the Market Ensemble. These capacities to mobilize are highly developed, but the Market Ensemble is problematic. American ensembles of consumption-production are too bound together with liberal versions of the Market Ensemble and there is a need to reduce the tendency to Taylorize. There is an interesting issue. Can and how might the producers in the Industrial Ensemble harness the capacities of the Intellectual Resources and the Interpersonal Ensembles in socialized-dedicated and specialized-generic products? This is a problem for the American government because in the past they have consistently structured the rules of the game around the principles of the Market Ensemble.

Finally, Storper and Salais propose four elements that should be found in a fruitful, analytically-structured narrative:

- Realism and indeterminacy
- Prior collective experiences and emergent forms
- Historicity and endogeneity
- Diachronic analysis to explain path dependent phenomena.

This is an important contribution to understanding organizational innovation as hybrid networks of knowledge, power and technology.

Diffusion of Innovation and the Suppliers' Gaze 7

S elling innovations is an international business in which the global service class acts as the eye of power, transferring selective innovations around the world. This chapter aims to show the subtle linkages between the innovation diffusion perspective and the interests of the supply side in marketing innovations. Chapter 7 examines the apparent dominance of the suppliers relative to the users of innovations. We say apparent dominance because those wanting to promote innovations, especially their suppliers, have generated the most influential frameworks in the academy and in the media. Figure 7.1 locates the suppliers on the upper left corner of the quadrant marked by professionals, state apparatus and users with the pool of innovations in the centre. This relational configuration is always being influenced by the national and the global contexts (Chapters 5 and 6).

The frameworks constructed by Hagerstrand (1978) and Rogers have shaped the approach to process known as the diffusion of innovations. The perspective of Hagerstrand proved some of the basic ideas, especially the time–space packing and routines in organizational innovations. The single most influential perspective has been that of the diffusion of innovations which is associated with four decades of research and theory-making by Rogers. His seminal and extensive contribution is from the framework of communication theory and so emphasizes the role of social persuasion and the sensitivity of the user to the non-rational aspects of adoption and innovation. Rogers' approach, which is very close to themes in marketing, involves making innovations appear and be fashionable for their adopters. One feature of the supplier approach is to black box innovations to shape perceptions of the innovation. This includes the anticipation of user resistances. A good example of this black-boxing is the approach known as Business Process Re-engineering (BPR). BPR can be understood within a Schumpeterian framework as an agent of creative destruction removing low-value knowledge.

The early contributions from the research programmes in the modern period (see Chapter 2) had a strong bias towards the quick adoption of all innovations. In fact, there are many problems with this pro-innovation bias and these are examined below. The main pro-innovation biases can be noted. Containing the biases is difficult because there are powerful tendencies that equate the rejection of innovations and opposition as discredited acts of Luddism (Grint and Woolgar 1998).

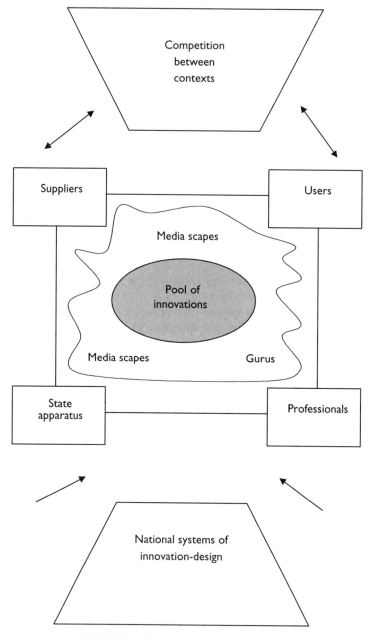

FIGURE 7.1   Pool of innovations framework

Recent studies of corporate and managerial innovations suggest that many innovations have a period of being fashionable in the media. So, are the media scapes creating theatres of consumption (see Chapter 6) and, if so, are the consumers gullible? A growing number of studies have profiled the rise and fall of innovations. These studies may be interpreted as suggesting that the fashions may merely be fads and that the choice of the innovation is inefficient.

Hagerstrand and space–time trajectories

Hagerstrand and Rogers provide the foundational thinking, conceptualization and directions to the notion that innovations spread from an original setting to other places. They each sought to conceptualize patterns in the flow of innovations and to explore some of the barriers to the successful 'selling' of innovations. Their approach is known as the 'diffusion of innovations' perspective. Although not intending to, they also provided much of the feisty rhetoric of the pro-innovation tendencies that became so problematic.

Hagerstrand's (1978) contribution can be examined under two headings:

- The basic iconography of innovation patterns
- The issue of resistance arising from the clash between routine actions and the time–space features of innovations.

First, Hagerstrand's (1978) early studies modelled the diffusion of innovations to detect their spatial patterning over time. Three findings have become highly influential icons and these are shown in Figure 7.2:

- The 'S' shaped diffusion curve
- The neighbourhood effect
- The hierarchy of places effect.

The 'S' shaped curve only examines successful innovations and aims to show the life process of successful innovations over time on the horizontal scale. The vertical scale shows the proportion of the final total population who had adopted the innovation against particular moments on the linear time scale. New innovations take about one-quarter of their life-span to become known and then in the next half of the life-span they spread rapidly until saturation of the population is achieved. The neighbourhood and hierarchy effects shape the curve of diffusion.

The early accounts of these three features emphasize that users are 'imitating' each other as a result of a simple, rational learning process. This neo-rational view of the supplier is still very powerful in marketing texts. These three icons of successful innovations have all too often been treated as the normal situation for all innovations, but the problem here is that we know that many new products and services fail after their launch. The exact proportion is probably under-reported but we have estimates of 50–65 per cent from highly regarded academics in marketing. We might expect organizational innovations to have a worse success rate. Therefore the three very seductive icons should be used with care.

129

(a) 'S'-shaped curve

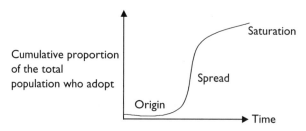

(b) Hierarchy effect

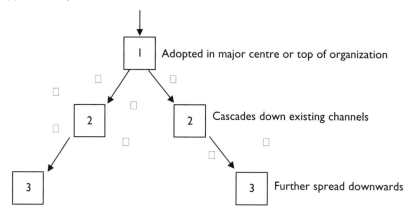

(c) Neighbourhood effect

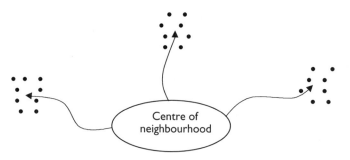

FIGURE 7.2   Icons

Second, Hagerstrand became keenly aware that potential users might not adopt innovations when he became involved in the application of operations research techniques to the problem of scheduling bus routes and timetables in Sweden. The Swedes seemed very reluctant to use the buses. Why? In a series of clever social analyses, Hagerstrand incorporated the notion that routines are of more influence than rational decision-making. Hagerstrand theorized that families and individuals ordered their everyday life around key events in their day. Their charting places a vertical scale of a 24-hour day with a horizontal map of the space occupied by the actors under observation. The time–space trajectories illustrate the daily pattern with key events and the chunks of time spent in various activities. This is an activity approach in which there are significant assumptions about the recursiveness of everyday life (see Clark 1975, 2000).[1] Those assumptions influenced Giddens' account of structuration.

Hagerstrand developed the approach of time–space geography with a whole group of co-workers. Their approach was the focus of a knowledge-pooling conference (Carlstein et al. 1978a, 1978b). One important element was the notion that there is *time–space packing* for everyone and that new activities have to fit into an existing 'portfolio' of activities that were constructed inter-subjectively. Hagerstrand constructed charts using the 24-hour day and a series of space-lines to display individual trajectories. Hundreds of studies have followed this format and they display the daily, weekly and other social rhythms of everyday life at work and everywhere else. The notion of time–space packing suggests that some innovations might be perceived as assisting the tightness of packing. For example, the creation and distribution of refrigerated packaged convenience foods exchanges money to buy the packages for savings in time. The savings in time include not having to buy and prepare vegetables, not having to use multiple kinds of cooking equipment and robotizing the short time required in the microwave. Likewise, Ikea play a major role in space-saving activities through the design and packaging of the self-build furniture and also because of its compactness and versatility in the home.

Framing innovation-diffusion: Rogers

Rogers' seminal research programme

Rogers' contribution started with an exemplary account and framework for undertaking the practice of innovation and for identifying researchable problems (Rogers 1962). From that seminal text Rogers expanded the analytic coverage and continually revised the original text (1962, 1976, 1983, 1995). Rogers' models of the diffusion of innovation have synthesized earlier research on central themes in diffusion and marketing and have been upgraded over the past four decades. Rogers sought to discover which processes and which contextual factors affect the rates of adoption and rejection. This led into differentiating between early and late adopters

---

1   There is a detailed account of 'recurrent action patterns' in Clark (2000: 94–95, 230–248, 269–270) and in the recent approaches of some evolutionary economists.

in order to explain the 'S'-shaped diffusion curve. Rogers also gave extensive attention to the role of networks (1962, 1983), while also being critical of those who treated networks as the 'turbo-charger' of innovation adoption (1995).

The initial models provided an heuristic knowledge for the suppliers to target the users of their products. The models covered the period from the commercialization of inventions to their sale or rejection by the users (purchasers). Initially the models did not explore what the users did with the innovations. There are thousands of research studies citing these models and they constitute the most influential account of diffusion. The early models were strongly inscribed with the background assumption that we should be pro-innovation and that those who opposed innovation were laggards (Rogers 1962) and Luddites (Grint and Woolgar 1998). Rogers recognized the pro-innovation bias and continually sought to revise those background assumptions (1983, 1995).

Rogers' original and seminal collection of models successfully assembled, compacted and blended knowledge from marketing, communication studies and especially from the Agricultural Extension Agency in the USA. In the USA the problematization (as PIEM) of the diffusion of innovations had been constantly undertaken. This area illustrates the exceptional American capacities in the commodification of knowledge.

The fields of diffusion and marketing are two interwoven knowledges that are designed to raise and justify revenue from consumers and from the state. Major players in the USA are passionately interested in diffusion and marketing: drug companies, scientists in agriculture, the retail trade, manufacturers. Communication studies are a highly regarded contribution because of their focus upon the consumption psychology of the spender. Periodic episodes of pooling, assembling and reflexively monitoring (editing) the knowledge already compacted into heuristic models occurred in each decade. Rogers (1962) was based on the pooling and editing events in the late 1950s when there was a strong input from research on farmers.

Bias to centre–periphery models

Clark (1987, 1997) contends that in the USA the problem of constructing best practice lists of 'how to do it' is both an everyday discourse on useful knowledge and is a long-standing institutional capacity to commodify knowledge. By commodifying knowledge we mean the capability to create commercially viable activities in which users would pay experts for their advice, services and products. The tendency was particularly strong in two key areas: the military and agriculture.

In the nineteenth century, agricultural production in all its aspects was the stimulus to enormous American exports to Europe and to the development of transport systems (e.g. railways: Chandler 1977) and transit hubs (e.g. Chicago: Cronon 1991). From an early stage there emerged specialist centres to accumulate knowledge. In some cases the centres undertook generic research into agriculture through experimentation. They became known as the Agricultural Extension Service. Its centres sought to ensure that their scientific output was perceived to be 'best practice' and that farmers adopted the best practice. By the mid-twentieth century a number of states possessed large-scale, quasi-independent research institutes attached to land-grant universities. Those centres were funded at the state level. Their performance was evaluated politically before new funds were allocated. The evaluations led the centres to fund social science research on the process of adoption. In the late 1950s there was a burst of activity in pooling knowledge.

Rogers' highly cited book is a grand narrative about how scientific research on diffusion could promote the adoption of innovations. It provided both an overarching view and also scripts for the different stakeholders involved in adoption. Because of the grounding of the frameworks in the experiences of the Agricultural Extension Service, the grand narrative presented the centre–periphery model. This suggested, quite plausibly, that the centre was politically independent and scientific. So, the book was an exemplar of the modern movement with experts from the centres applying neutral, true, objective knowledge to the real world to achieve progress (Clark 2000). Consequently, the innovation supplier did not appear in the key diagrams. However, the key diagrams, like those of Hagerstrand, attained an iconic status in the new field of innovation studies.

Unintentionally, the example of the Agricultural Extension Service obfuscated the role of the supplier, yet 'placed' the supplier in the centre of the analytic field. Because the supplier was presumed to know best practice, Rogers' narrative treated the central problematic as one of broadcasting the news about new innovations. It is presumed that the farmers should adopt and not resist because the innovations are thoroughly researched and designed for them and for America. The user role is largely passive.

The specific model of innovation produced by Rogers promoted the centre over the periphery and expressed a *broadcast-receiver view* of the supplier–user relationship. Moreover, key elements of the knowledge codified and transposed into 'innovation-diffusion' were actually transposed from the field of marketing. For example, at the core of Rogers' (1962) original five-stage model of the user's decision process was a mnemonic known as 'AIETA': awareness, interest, evaluation, trial and adoption. AIETA was mentioned in the marketing encyclopaedias of the late 1920s. So, it is not surprising that this centre-oriented model was re-incorporated into marketing texts for students and became the framework for research and practice by which the drug industry sought to influence doctors. In this case the interests of the supplier were more immediately obvious than were those of the Agricultural Extension Agency. The hidden influences of the Agricultural Extension Agency on the diffusion of innovation model are considerable.

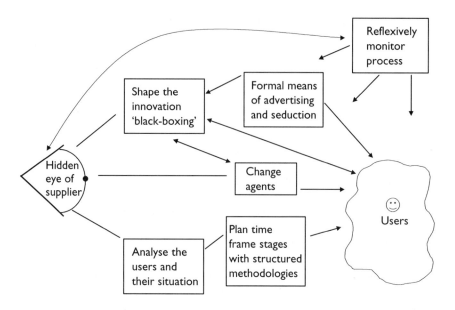

FIGURE 7.3 Suppliers' hidden eye of power

Two-step communication

Centre–periphery models adopted the *two-step theory of communication* which is shown in Figure 7.3. The communication theory presumed that individuals were influenced by their peers and by certain features of the innovation. The application of the theory was based on the notion that target populations possessed distinctive characteristics that were *ante- cedent* to the arrival of innovation and that those characteristics would distinguish communities that adopted early from those who adopted later. Implicitly the two-step model presumed that there were network effects, as reported by Hagerstrand (1978). The application of the two-step approach acknowledged these antecedent features. The first step should consist of the massive supply of documentary information infused with features designed to enrol the targets. This was mainly in the form of textual information but gradually came to include more explicit advertising and to be interwoven with research on fashions (Lears 1994). The formal media were used to target the specific users (e.g. farmers). Second, the target community was analysed by *change agents* who aimed to locate those members who were highly regarded and likely to be early adopters of the innovation. The change agents would then engage with those targets to reinforce their decision and to persuade them to engage in trial adoptions. The change agents presumed that the adopters would be successful and that the opinion leaders would start to influence their peer group even before the outcomes were known. Change agents were given checklists of charac- teristics to look for in the target community and in its key personnel. They were warned to stay away from those persons known to be eccentric or to have a poor record of output. The change agents were expected to become knowledgeable about particular settings.

Black-boxing innovations

The centre is encouraged to design their innovations to enrol the perceptions of the potential innovators. Five dimensions of perception are highlighted: offering *relative advantage* to the adopter; being *compatible in values*; *minimizing complexity*; *dividing the adoption process into stages*; and showing results that can be used to *communicate performance*. These are a mixture of the modernist concern with reason and the marketing awareness of fashion, self-esteem, social status and seduction. The five features include the following:

(1) The innovation should be perceived as giving the *adopter relative advantages* in financial terms (e.g. better use of time and space) and in terms of status in the community. Convenience foods offer obvious advantages in time use because they are quick to prepare and do not involve the ancillary tasks of buying the ingredients and disposing of the waste.

(2) The innovation characteristics should highlight areas of *compatibility* to existing values. For example, Time-based Competition appeals to those who value time and Business Process Re-engineering has an appeal to those who believe they can be decisive and cost-conscious.

(3) The *complexity* of the innovation should be *minimized* by careful pre-design of functions and by careful cueing of the adopter's perception.

(4) It is desirable for the adopter to be able to engage in actual and imagined *small-scale experiments*. If possible the adoption process should be divided into steps. For example, a new seed for farmers could be planted in one acre as an experiment. Farmers might imagine this safe step and then conclude that they can move forward to more large-scale adoption. In the case of many systemic information technology innovations the suppliers can offer the adopter the possibility of buying small pieces and then gradually building up the system. This strategy is used extensively with SCITS (see Chapter 2).

(5) The *outcomes* of adopting the innovations should be presented in *quantified* form. For example, farmers who purchase a new field irrigation system should be able to identify performance dimensions they can monitor (e.g. effectiveness of time of supply to the field).

These five features indicate that the suppliers should black box the innovation to transform any fuzziness (Clark 1975) into perceived certainties and confidence about adoption.

In the original models the innovation was conceptualized as stable through time. Rogers clarified this by emphasizing that the users might

135

re-invent the innovation after adoption by their usage. The idea of re-invention is an important analytic opening, but it will have to be developed more fully in Chapter 8. For the moment we should note that the diffusion process tends to stop at the adoption stage when the innovation has been purchased. Stopping the analysis at the adoption moment was not problematic in the centre-broadcast model because there were many studies evaluating how innovations were used. Academics found the clear data provided by the date of purchase very useful in the modern research programme (see Chapter 2).

Firms as users

Rogers and Rogers (1976) extended the analysis of the change agent into the field of strategic planned change. Zaltman and Duncan (1977) found the initiation/implementation dichotomy gave models a tonality of certainty even when dealing with uncertainty in the models. However, Eveland (1979) and others noted that although the same shape of innovation was adopted in similar settings, the actual usage could vary enormously and the supplier had not envisaged some uses. Their work drew attention to the events after adoption and also to marked differences between the individual adopted, such as the farmer and the local medical practitioner recommending drugs from the large-scale corporation. Rogers and Rogers (1976) added further models to inform the supplier about the user, especially the firm as a user. Clark and Staunton (1989) use those models to develop a decision episode framework (see Chapter 8).

Pro-innovation bias

The pro-innovation bias (Clark 1987) is a discourse saying:

- Innovations should be adopted
- As quickly as possible
- In the format recommended by the suppliers
- Innovations tend to be fixed in their constitution and uses
- Innovations are essentially efficient
- Adoption will probably be successful
- Adoption improves the effectiveness of the firm
- The suppliers distributing the innovation are politically neutral and rational
- The user is passive awaiting implementation
- There is a one best way of singular best practice
- Adoption contains stages in a linear sequence.

The pro-innovation bias is so pervasive in the media that even articles prefaced by its recognition are guilty of its promotion. It may be noted that in diffusion studies the analysis is micro-level and ignores the multiplicity of levels in the embedding context. Also, the pro-innovation bias tends to prioritize discursive penetration and strategic agency over actionable knowledge.

Clark and Staunton (1989) contend that the discourse of public accounts of innovation is riddled and muddied by the pro-innovation bias. The bias is especially deep in the first and second research programmes examined in Chapter 2. The bias is so pervasive that even essays prefaced by its recognition are included in its promotion. Therefore it is essential to recognize the political processes of innovation and their *contested, confrontational* features (see Chapter 4). The political context of all innovation including network innovations, has been conflated, eviscerated and suppressed. It is the pro-innovation bias that has influenced theory-building.

Opposition to innovations is ordinary and extensively implicates management. Pettigrew (1985), in a very detailed examination of the decade-long introduction of Organization Development (OD) in the five divisions of the Imperial Chemical Industries (ICI), shows that four of the divisions actively opposed the introduction of OD. Only within the Agricultural Division did OD gain access to the strategic decision arenas and then budget cutbacks in the ailing firm undermined developments. Despite these rejections, the narrative invokes OD rhetoric to explain success in the Agriculture Division as the felt need for change, support of key individuals, enrolling opposition and use of outside experts.

The counter-point to the pro-innovation bias is that:

- Much if not most innovation in organizations occurs through alterations to the population of organizations through the exit of surviving firms and entrance of new firms. More than four-fifths of the firms founded two decades ago have closed down and most of the remainder are still mini-firms
- Most innovations in new products and services fail even though the innovating firm survives
- Dis-order relative to economic efficiency is the typical situation
- The inertial capacities of the existing capabilities of surviving firms are powerful constraints through time
- Many surviving organizations either change their parenting and/or engage in strategic alliances to protect their incapacity
- Its own management may not reflexively know the managerial expertise of a successful firm. Corporate management should use the notion of Penrosian learning (Penrose 1959) to reveal the knowledge of the firm and to reveal how that knowledge is embedded in particular contexts.

Unpacking the politicality of organizations and their contexts represents an important way of advancing understanding. Grint and Woolgar (1998) provide an extensive account of the politics of innovation, especially contest and resistance. Gruneau's (1999) historical essay on Canadian sport reveals the everyday life of contested innovation. Critical political economy succinctly expresses alternatives (e.g. Noble 1984, 1998). How might the political metaphor be improved? Frost and Egri (1994) propose a theory of surface and deep politics to replace the political metaphor of pluralistic, self-interested, non-rational action with the role of humans as heroes, victims and villains. They raise problems with approaches such as surface and deep politics when these are focused only on means to control the innovation, that is the implementation subscript of the pro-innovation bias. Frost and Egri propose to situate the deep/surface framework within a focus upon means and ends in a longitudinal coupling of social and technical interests. Theirs is a search for patterns within and between interrelated innovation situations that recognize system complexity, requisite variety and chance. They give an invaluable stress on having a time span that embraces past, present and future.

Fashion

Figure 7.1 locates innovations in the centre surrounded by gurus and media scapes. Do process innovations appear in life cycles of emergence, spread and decline, each lasting one or more decades? And, if so, what are the implications? Does useful learning occur or should the efficiency hypothesis be rejected? This section focuses on the examination of fashion (see Chapter 6).

Rogers (1962/1995) acknowledges the role of fashion in the two-step approach whereby suppliers start by using advertising and other structured forms to trigger interest in potential buyers in those process innovations they wish to merchandise. They later follow this up with personal contacts with potential buyers. The life cycle corresponds to the theory of the 'S'-shaped curve of diffusion and many examples are now available depicting the life cycle of process innovations. Pascale (1990: 20) listed 27 examples between 1950 and 1977. Similar lists have been compiled for every functional area of management (e.g. Scarbrough and Swan 2001). So we must conclude that in every decade there are several score of process innovations being used and being rejected. Rogers' approach presumes that innovations are chosen and adopted for social reasons (e.g. esteem, status), and yet overall there is the opportunity for higher effectiveness.

Abrahamson and colleagues[2] have concluded from a series of studies done in the USA that there are fashion-like life cycles for some process innovations, but these are the exception. *Most process innovations emerge, are de-selected, disappear and are forgotten.* We shall examine Abrahamson's contribution to understanding fashion, starting with his core definition (1996):

---

2   The contribution of Abrahamson opens up different angles from those proposed by Swan, Newell, Robertson and Scarbrough.

- Fashions are relatively transitory
- Fashions are collective beliefs assembled and packaged rhetorically around symbolic labels (e.g. Business Process Re-engineering)
- Fashions are dominated by management fashion-setters (e.g. gurus, consultants, media)
- Management fashion-setters continuously redefine those collective beliefs, downgrading failing labels and re-positioning new labels. The fashion-setters surf from one fashion to the next
- Rational progress and performance gaps are closed because the 'solution discourse' is feasible.

The implication is not that managers are simply gullible, but rather that their engagement with process innovations has to penetrate a seductive discourse (Abrahamson and Fairchild 1999: 727). The suppliers invest heavily in a strong discourse of promoting techniques and rhetoric. They suggest genres of techniques so that the genre can survive while specific process innovations disappear. Genres are collections of interrelated innovations. For example, the genre of Scientific Management contained a number of distinct areas and techniques. Much of the genre is re-blended in Business Process Re-engineering. Discourse influences diffusion by developing new categories (e.g. Just-in-Time) and by revivifying old categories. For example, recent attention of SCITS is focused upon 'flows'. So, publicity material highlights the possibility that the flows of a hospital can be analogous to the flows into, through and out of supermarkets and that patients will be the beneficiaries. Fashion-setters construct the 'appropriate rhetoric' with the promise of both feasibility and performance improvements from using the process innovation. The performance discourse highlights the dramatic gap and the solution discourse presents the new process as the rational solution. However, management fashions are neither cosmetic nor trivial and are more than simply an aesthetic. Fashions shape the techniques and prime the users' capacity to problematize and mobilize networks within the firm (Abrahamson 1996).

There are markets for process innovations with well-organized suppliers situated within the media scapes shown earlier in Figure 7.1. Within the market there are niches for management fashions and because these have a finite carrying capacity the suppliers are extremely active.

The life cycle is conceptualized in terms of the VSR framework of variation–selection–retention introduced by Weick (1969). The appearance of a new variant (e.g. *Kanban*) is followed by a period of being selected-out (typical) or selected-in (rare) within the market. If selected-in, then this might lead to the institutionalization of the new process (e.g. *Kanban*) in the firm. The latter is the retention mechanism in the VSR framework. Abrahamson is sceptical of researchers who find the 'S'-shaped life cycle model everywhere because so many process innovations fail (see Rogers 1995: 104–114). The VSR framework has been applied to a wide scope of databases in an imaginative use of proxy measures of the processes. The conclusion is that process innovations rise and fall as transient, weakly institutionalized forms whose retention and institutionalization are the exception. Clearly, Abrahamson offers a different and in many ways more persuasive analysis of the 'life and death' of process innovations than does the early work of the new institutional school of DiMaggio and Powell for the narrow and somewhat misleading focus upon retention. The aim should become to alter the agenda towards a more searching examination of the claims made about the supply–user processes. Abrahamson does, however, confirm the claim that process innovations are constellations of uncertainty and he does amplify the role of the supply side in a way that suggests revisions to the isomorphism hypothesis.

The de-coupling of the genre discourse from the specific technique and from any organizational changes that might occur in the user-firm is a useful one, yet needs carrying further. Users have to penetrate the media scapes, discourse and fuzziness that surrounds organizational innovations (Clark 1975). One of the earliest studies of the use of work-study, a subset within the scientific management genre, found that a keen and willing management team interpreted work study quite differently from their consultants. Since then there have been too few similar studies of the users. The few completed studies do reveal gaps of comprehension between suppliers and users. Moreover, through experience, suppliers have become more adept in their construction of discourse and in *learning from the user* (Von Hippel 1988).

Kieser's (1997: 52–54, 71) approach is heavily centred on the claim that consultants insert a fusion of rhetoric and technique into the decision arenas from which users take their innovations. In the arenas approach there is struggle between competing games with one set of players having a very powerful position. The consultants and their rhetorics hold the most powerful position. This leads Kieser to argue that management techniques are not available in pure form without rhetoric because consultants sell their wares through rhetoric. However, as Abrahamson shows, there is a de-coupling between the discourse of a genre and specific techniques. That uncoupling is very clear in the case of SCITS (see Chapters 8 and 9).

The position of the media was introduced in Figure 7.1. Faust (1999) also contends that the mass media are one of a set of loosely connected arenas providing potentials for both homogeneity and diversity. This interpretation will be applied here. The mass and quality media tend to occupy a gate-keeping role in blocking or facilitating the flow of fads and fashions and they may act as a primary regulator of innovation (Abrahamson 1996; Czarniawska and Jorges 1996). Their role is mainly in stimulating and confirming that a particular fad or fashion has entered on to the agenda for consideration. Frequently, the media receive extensive public relations literature from the consultancies. Influential newspapers such as the *Financial Times* and the *Wall Street Journal* and weekly specialist publications such as *The Economist* are prime targets and sophisticated translators. The mass media are used by a plurality of interest groups: employers associations, unions, professional associations, the government and sectors of industry. Parts of the mass media have strong links to another arena containing management consultants, book publishers, gurus, academics and entrepreneurial executives.

The isomorphism hypothesis, along with similar theories of homogenization, tends to emphasize the opinion-shaping role of the mass media. However, the existence of multiple arenas, coupled with the examination of the specific role of the professional associations involved in logistics, inventory and production control by Swan, Newell and Robertson (see footnote 2 on page 138) suggests that the arenas might be rather scattered, locally focused and with structural holes (R.S. Burt 1992).

The position in the mass media of the new managerialism associated with Anglo-American approaches to positive political economy suggests that it is background assumptions that are most circulated through the media. The discourse of media tends to contain stories with particular genres. In Britain, the mass media have relayed and built upon the audit society (Power 1994) and have successfully dramatized the league tables on hospital performance (e.g. death rates). This is much more evident than the quick mentions of specific fashions such as Business Process Reengineering, the Learning Organization or Knowledge Management (Scarbrough and Swan 2001). The discourse constituting role of the media in background assumptions is probably more significant than the ephemera of passing fashions among managers.

Templates and structured methodologies

How are organizational innovations commodified, diffused and marketed to firms? Both the isomorphism school and also Chandler (1962, 1977) impute mechanisms, but don't conceptualize the processes. The diffusion perspective of Rogers does highlight a number of important features but leaves the key area of core processes and mechanisms undeveloped. The isomorphism approach implicitly follows the same analytic principles as Rogers and therefore also fails to clarify the central process. Hasselbladh and Kallinikos (2000) contend that the theoretical formulation of the isomorphism hypothesis is too idealistic and does not include the *discursive and codified means* by which rationalized systems are made stable, durable and portable. There is no account of the means by which the domain action encapsulated in the organizational innovation is conceived as rules of conduct and performance principles. Nor is there an account of how devices of control are developed and forms of actorhood constituted (ibid. 2000: 701). A theory of institutionalization should conceptualize and explain the organizational processes through which organizational objects, procedures and roles develop and become embedded (ibid. 2000: 703).

The concept of template is used to conceptualize an ideal that influences the construction of structured methodologies (Clark 1987). The concept of the template is a subset of structuration theory and is a special case of situated practice in structuration and habitus (Bourdieu 1977). Templates are generalized cognitive frameworks or typifications deployed by some groups to impose an orientation upon action and to give those actions legitimacy and meaning in a specific domain (Harré 1979). Templates cue and are cued by affect and behaviour. They consist of a known *generative pattern* with rules and actions from which replicable sequences of action have been derived. Universal examples are found from sports and also from the templates for erecting boxed kits of furniture. The generative pattern is sufficiently articulated, embodied in inscriptions (e.g. drawn shapes) and made portable through competent bearers as they and their constituency require. If a template exists, then some sequences of action are highly articulated and can be extended through time and across space.

The concept of template incorporates the social constructionist perspective and clarifies one key aspect of the isomorphism approach. Templates vary in both their articulation and in their public ownership. For example, the McDonald's trading and franchise system is an array of templates that are highly articulated within the firm. In the McDonald's University, the various templates are presented and rehearsed at special learning sites. From there the new bearers carry the template into new locations. Moreover, some aspects of the template are readily observable to the outside analyst. It may be noted that 'Pret à Manger' has many similar template features although these have been applied to a more seemingly customized range of consumables.

Some templates are only slightly articulated. This was the case for rugby union in the 1870s when American college males experimented with the sport. Americans found rugby contained features that were not exciting or appealing to their sense of strategic manhood and therefore introduced key, pivotal modifications – the 'down' – that led to the emergence of a new template. Later that new template was further elaborated with the introduction of the forward throw. Most organizational innovations are only slightly articulated into templates.

One of the features of templates is that their construction in one setting may inhibit transfers to other settings. Lillrank (1995) observes that organizational innovations do not transfer readily in their original packaging. He identifies three features of structured methodologies that influence the effectiveness of their transfer, specifically between nations:

- *Management principles* and strategy implicating the success factors and organizational building blocks. For example, Toyota principles typically emphasize quality, flexibility, consumer satisfaction and long-term growth.
- *Organizational vehicles and structures* embodying the strategy. These are complementary to ideals and strategy and are highly contextual because they include local incentive structures, labour market conditions, cultural dispositions and governance.
- *Generic techniques and tools*. These tend to be very specific and to require a specific infrastructure.

142

Lillrank contends that the transfer of Japanese Quality Control Circles to the west largely failed because the western importers failed to understand either the systemic articulation of these core features or the need to adapt and appropriate those elements required and fitted to the new contexts (Clark 1987: 153f.). This complex organizational innovation was pushed through transfer channels that were only capable of handling simple, low-level abstraction. Moreover, the western receivers possessed different principles, organizational vehicles and techniques. Some of the low-context tools survived but the organizational infrastructure failed. Later, however, Time-based Competition (TBC) was transferred successfully by internationally experienced consulting firms. TBC is the pursuit of competitive advantage time compression and commodification (e.g. reducing cycle time). The aim is that deliveries can be made sooner and that the time spans of demand forecasts are more accurate. These reduce capital costs and work in progress. Because precise time has some symbolic similarities to money and is already part of the profession of industrial engineering (and cognate occupational groups) its transfer was much simpler than for Quality Control Circles. The development of TBC started in the late 1970s when certain Japanese firms in the automobile industry were observed to be using variations of western practices. International consulting firms, already in possession of basic time management skills, were soon observing and appropriating the strategy, its embodiment in organizational vehicles and the various tools. But the quality aspects were disconnected from the time frames. The structured methodologies in the consulting packages focused tightly on time frames. TBC provided a stimulus to the development of western awareness of Concurrent Engineering for improving product development (see Clark and Fujimoto 1991). TBC also articulated with Business Process Re-engineering and its core features were unbundled and redistributed.

Hasselbladh and Kallinikos (2000) also develop a reformulation of the limits in existing approaches through a Foucauldian framework applied to processes and forms of objectification and actorhood. The objectification and the processes of institutionalization involve and can be examined in terms of the general approach of Foucault. *Four social processes* are identified:

(1) *Ideals* (goals, social intentions). This is the semantic delimitation of closure of ideals that are accomplished by specific verbal and written strategies. Ideals are developed into social relationships and causal models.
(2) *Discourses* (social roles and performance principles). Discourse primarily conducted in those written forms that become both sedimented and also the basis of actions. Latour and Callon have already primed this aspect.
(3) *Techniques of control* (taxonomies, codification, systems of measurement). Techniques of control are usually in the form of accounting systems based on taxonomies that codify and measure.

(4)   These three become *articulated within domains of activity and action.*
      Articulation demands the support of other groups through some form
      of enrolment (see Chapter 4).

The outcome is a variant of a structured methodology containing a
sequence of instrumented modules that are in the form of metrics. There
are well-defined rules and procedures based on standardized symbolic
schemes. Once institutionalized, the objects are relatively durable, can be
communicated and therefore have the potential to be reproduced in other
settings.

Structured methodologies vary in their codification and this is influenced
by the discourse surrounding the specific innovation. Clark (1975) contends
that organizational innovations vary in their fuzziness as perceived by
particular professions (low fuzziness) compared to the users (high
fuzziness). Fuzziness is a core feature of the Decision Episode Framework
(Clark et al. 1992–93). Some innovations will be highly contested. For
example, Djelic (1998) claims that American innovations for transforming
European corporate capitalism were differently contested in France (high),
Germany (less) and Italy (high) in the period 1945–70. The British case is
particularly interesting for the levels of covert resistance in the engineering
industries (Zeitlin 1995) and for the degree of overall American penetration
and Americanization through population ecology effects in the replacement
of British by American-owned facilities (Mathias and Bjarnar 1998).

Objectification is part of dual process and requires translation through
knowledgeable and influential agents into specifics on arrival at some
context. This transition from explicit knowledge into local knowing
communities of practice (e.g. assemblages) is central to the organizational
innovation. How are structured methodologies actualized in particular
settings? Neither the diffusion perspective of Rogers nor the new institutionalism of DiMaggio and Powell or Nonaka and Takeuchi quite
unpacks this process.

Objectified templates and structured methodologies may be commercially valuable. This leads to the black-boxing of the template. In this
process, the detailed workings of the template are obscured from view and
replaced by some symbolic representation that implies their capacity to
deliver. This capacity may be more rhetorical than substantial. Later we
shall find that some software systems for strategic co-ordination were made
to look more operational than in practice.

Next

Figure 7.1 located the role of the suppliers and the media in shaping
organizational innovations. Now it is essential to examine the pool of
innovations in the centre of Figure 7.1.

Organization Process Technologies and Hybrid Networks 8

P rocess innovations are part brandscape, part technoscape and part material configurations. Also, they possess a degree of autonomy from those who claim to supply them (Chapter 7) and from other actors like media, gurus and academics who endlessly proselytize about them. That relational configuration was expressed in Figure 7.1 with a pool of innovations surrounded by suppliers, users, professional associations and the state apparatus. Both national systems of innovation-design (Chapter 10) and global competition between contexts (Chapter 5) also shape the size and scope of any innovation pool. This chapter selectively positions concepts and propositions from previous research and the earlier chapters.[1]

Organizational innovations and process technologies are socially constructed 'facts' in a relational configuration involving many different actors. Different relevant groups will interpret the process technology according to their values, power and interests. Therefore the analogy of a template (Chapter 7) should be used in conjunction with interpretative flexibility (see shortly). Templates can vary from being precise, explicit and complete with clear examples and prototypes (e.g. the franchised outlet of a McDonald's) to being very imprecise, fuzzy, heavily contextualized and without prototypes (e.g. German corporate practices at Volkswagen).

Process innovations are dispersed in technoscapes (Chapter 6), are equivocal (Weick 2001) and are not tightly objectified templates. Moreover, their constitution implicates a configuration of actors, both human and non-human. The technoscapes of innovations contain a range of substitutable alternatives rather than singular best practice. Process innovations are deeply contextual, yet are translated into non-situated entities. Consequently the users and suppliers experience considerable difficulty in establishing an organizational innovation in a new setting. Scarbrough (1995) uses the language of prisoners, hostages and black-boxes to conceptualize the interactions between suppliers and users.

The framework shown in Figure 8.1 demonstrates a configuration with three major clusters of players: the innovations pool, the supply side and the users. This chapter brackets the influences of the national innovation-design systems and the international political economy. They are always present and are re-examined in Chapter 10. The pool of innovations is

---

1 This refines the propositions in Clark (1987: Ch. 6) and Clark and Staunton (1989).

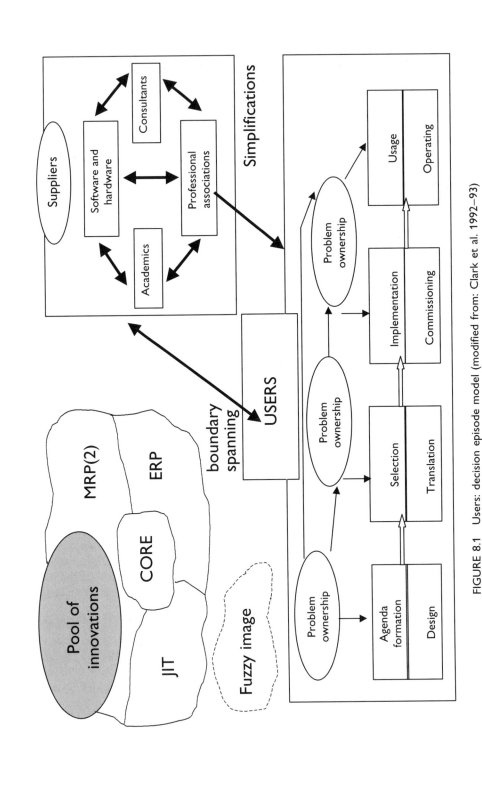

FIGURE 8.1   Users: decision episode model (modified from: Clark et al. 1992–93)

created by many players, including the suppliers and users, but also the media and academics. Pools are constituted to contain innovations that appear to address similar situations and problems. For any given use there are likely to be a multiplicity of innovations, as shown in Figure 8.1. However, the pool might exclude some innovations that dovetail together. The 'pool of innovations' is the focus of this chapter.

Organizational innovations possess a degree of autonomy. Moreover, most are not publicly accessible except through invitation. The exception is of sporting innovations that are in the public domain although many of their configurational features are hidden in rehearsals. With sporting innovations there is a high degree of spying on rival teams at their rehearsals. In the absence of public access, social constructions are made available as 'scapes' arising from the actions of the media (e.g. television, journalism) in producing inscriptions (e.g. texts and videos). Later in the chapter we take the specific example of strategic co-ordination and control innovations.

Dynamic contingent configurations

In this section five aspects of the pool of innovations are examined. We start with a set of guiding propositions drawn from earlier studies:

(1) Unbundling and hybridization.
(2) Interpretative flexibility.
(3) Design hierarchy.
(4) Innovation variants and the supplier/user junction of selection.
(5) Retention as appropriation.

First, many organizational innovations are unbundled and transformed into hybrids because they are:

- *Blendings* of elements shaped into commodifiable bundles by particular groups in specific contexts
- *Relational materiality* in which social elements and material elements are interwoven into networks containing heterogeneous combinations
- The interweaving of social elements with the natural context (e.g. weather, geomorphology) and artificial elements such as technology and draw materials into *hybrid networks*
- The architecture of the elements in terms of the configuration and the *tightness/looseness* of the bundling. These features are illustrated later in the chapter by examining a software based co-ordination system we refer to as SCITS. This contains many modules arranged on common platforms. However, that viewpoint is from the suppliers', rather than users', perspective
- Should be examined in *dynamic evolution over long periods*, possibly of several decades, in order to understand the contingent dynamics of the configurations that assemble them

147

- Epochal innovations that arise because the previous situation contained a socially defined blockage – a *reverse salient* – which had to be removed through a design investment. Today we take bar coding so much for granted that it provides one means for calling the daily roll in schools as well as for the audit of supermarket shelves. Yet, bar coding was visioned in the 1960s long before its relational materiality could be delivered and even longer before its novelty as an organizational innovation was appreciated
- Longitudinal studies that have to avoid false linearity. Retroduction will show both *multi-linearity* (Bijker et al. 1987) and also many possible futures, including emergent outcomes that are not expected. American Football at its founding was an emergent outcome (Clark 1987). This means using concepts such as social trajectory and path dependency with considerable caution
- Innovations that often emerge as *unintended outcomes* of struggles between pre-existing organizational innovations
- Innovations that vary in the extent to which their *initial founding* conditions affect their subsequent evolvement
- Transformed by small pivotal alterations that alter the overall architecture of the configuration. For example, the introduction of the 'down' in American Football in the 1880s
- Subject to the influence of those suppliers who exaggerate the degree to which an innovation resolves the problems for which it is sold. This is known as rhetorical closure (Bijker et al. 1987)
- Subject to the role of users whose use of the innovation may conflict with the intentions of the original suppliers. For example, credit cards . . .'
- Re-moulded by the users' re-invention and re-definition of their core elements. For example, when British universities purchase Enterprise Resource Planning Systems (Pollock and Cornford 2000)
- Transformed by the actions of the many groups (see Bijker 1995) that surround their introduction. Organizational innovations are are embedded in specific contexts and are shaped by many contextual features and the multiplicity of social groups.

Second, the concept of a pool of innovations refers to variations around a genre.[2] The genre refers to commercial, computer-based, software packages that the suppliers claim process key informational aspects of strategic co-ordination. Variations can include different parts/wholes of the genre and the different versions produced by competing suppliers. Variations also include the evolution of different solution-problem packages. In the socio-evolutionary theory, the pool of innovations face two quite different obligatory passage points. One is whether the users recognize the novelty

2   Genres are broad typifications devised stylistically to highlight collections of attributes. For example, Yates and Orlikowski examine genres of corporate texts.

and relevance of their solution and can institutionalize that solution. This is retention. For example, there is clear evidence that within the British publicly owned National Health Service there are a variety of innovations that do not travel around the system. This is because the social capital contains low capacities to notice and retain the innovation. The pool of innovations should be shown to include those innovations that were unsuccessful. However, that is rarely done with organizational innovations. Another source of loss of variations arises at the conjunction between the suppliers and the users. Cowan (1987) uses the example of American iron wood-fire systems in the nineteenth century to show that the users selected-in some variations and not other variations. In socio-evolutionary theory the pool should be more inclusive than is shown in Figure 8.1.

Third, the exceptional situation for an organizational innovation is that it is reproduced in a highly similar way in many locations across many nations. An obvious example is global media sport, specifically the Olympic Games. In the Olympic Games, athletes compete in more than 40 sport genres that are based on detailed, internationally accepted rules. Likewise, some firms manage to establish their brands in very similar ways in many nations. McDonald's is a useful illustration because their con-sumer products and services contain contextual variations beyond the original American template of a decade ago. Moreover, the degree of customization has increased and scope has been extended to include mid-price sandwiches from the acquisition of the 'Pret à Manger' chain. The Olympic Games and McDonald's exemplify situations in which an inno-vation is reproduced in a very similar manner over space–time.

The more typical situation is, contrary to the claims for isomorphism, one where an organizational configuration possesses some similarities and varying degrees of difference. Various studies show that scientific manage-ment differed in its practices between sectors within a nation and between nations, for example, differences between the USA and the UK (Clark 1987; Guillen 1994). One way of expressing this issue is to compare some socially constructed objectification of an organizational innovation with variations around the world.

Clark (1987) maintains that the *unbundling* of innovations is almost the typical situation. This presumes the existence of a stated frame of reference. For example, American Football is a hybrid running game in which eleven players carry an oval ball. American Football is contingently shaped from association football (eleven players kicking) and rugby union (fifteen players carrying) with a small input from indigenous folk games. The 'down' was introduced in 1880 and became the pivotal, small change that altered the design configuration of the game, its relation to the consumers and its position in American culture. Clark and Staunton (1989) contend that the Japanese unbundled a stream of American corporate innovations by:

- selecting some features
- modifying some features
- adding in novel features

149

- combinations of selecting, modifying and adding
- using the original/hybrid innovation in different contexts.

Abo (1994), in an edited collection, elegantly and persuasively argues that American factories interested in Japanese innovations became hybrids with both American and Japanese features plus emergent features. Unbundling and the hybridization of innovations are typical outcomes for organizational innovations. This interpretation qualifies some of the implicit evolutionary trajectory inscribed in Chandler's seminal account of certain American innovations (Chandler 1962, 1977, 1990).

Fourth, the linkage between the pool of innovations and the supplier is shown as 'fuzzy'. The implication is that even when the innovationscape seems to possess clearly objectified genres, as with major sports (e.g. association football), these are always subject to the alignment of social constructions between all the diverse groups involved. Here the recent evolution of rugby union rules and practices (e.g. professionalization) reflects the influence of different groups. The multiplicity of groups that might be relevant influences on an innovation was illustrated with the bicycle (Bijker et al. 1987; Bijker 1995).

Fifth, the multiple interpretations and fuzziness surrounding objectified innovation pose a particular challenge to the notion of singular best practice as noted in Chapter 7. Therefore, comparing the effectiveness of a genre of innovations requires careful specification. Unpacking that implied scale is also the aim of Best's (1999) intriguing comparison of the evolution of the competitive advantage of system characteristics which was discussed in Chapter 4. Appropriation refers to the competitive advantage of the specific variant of an innovation that users ingest and can utilize. A simple scale of appropriation examines this issue (Clark 1987: 153f.).

Process technologies as network design and building

Technology studies in the two main research programmes examined in Chapter 2 concentrated upon profiling abstract dimensions and constructing typologies (e.g. Pugh and Hickson 1976). These tended to focus upon the process technology and upon the characteristics of the product–market mix (e.g. Woodward 1970). The informational technologies were neglected. In the third research programme (see Chapter 2), technologies are examined as organizational innovations. We shall examine the social construction of technology for its account of network design and building.

The emphasis upon the multiple, heterogeneous shaping of the *design state*[3] prises open certain limitations in previous frameworks (e.g. Bijker (1995), Callon (1986), Latour (1991) and Akrich (1992)). However, there

---

3 The reasons for using 'state' rather than 'stage' are explained in the next chapter.

has been a tendency in the social construction approach to privilege t
network building and also the durability of heterogeneous actor networ
Consequently, the role of the user becomes oversimplified and this proble.
has been the focus of recent revisions to actor network theory. Addition
ally, the builder/designer–user dichotomy requires revision beyond simple
notions of implementation (Clark 1972; Whipp and Clark 1986; Clark et
al. 1992–93). The builder–user dichotomy parallels the crude, rude, yet
persisting distinction between strategy and implementation. This section
suggests some basic concepts and an approach that should be part of our
eclectic theory.

First, heterogeneous actor network theory contends that because all
social relations are permeated with the non-social, both natural and artificial (e.g. technology), then it is misleading to conceptualize and explain
phenomena only in terms of the social. This is the relational materialism of
configurations. In the approach adopted here, the heterogeneous actor
networks are all under the same roof. But how are these configurations
constituted and under what conditions do they maintain/lose their shape?
One of the major influences is the relevant social groups around an
innovation/technology. Each group will have its own active construction of
the configuration and therefore there are as many configurations as there
are relevant social groups. Pollock and Cornford (2000) reconstructs the
groups around a generic computer system inserted into British universities
in the past decade to include consultants, internal programmers, university
clerks, the Estates Office, the Finance Office and many others. In addition,
there are the non-humans, as in the seminal vignette of how biologists re-
enrolled scallops with the social group of Brittany fishermen. So, variations
between the configurations represent their *interpretative flexibility*.

In the four moment framework of PIEM (see Chapter 3), the problematization and *intéressement* moment is one in which actors raise issues
about the identity of one another by implying to them that 'this is what
you want to be' and offering a different scenario. The moments of translation and the obligatory passage points are conceptualized as a cybernetic
alignment of interlocking activities in which actors know what to expect
(also Weick 2001). There is some black-boxing to make an innovation
work unproblematically (see Chapter 7). In the enrolment moment an
interrelated set of roles is defined, attributed and allocated to actors who
'accept' them. So, configurations contain many co-existing networks and
each network lends force that cascades enlistments into the configuration.
Scallops, for example, become enrolled through researchers' negotiations
and the scallops are transformed into intermediaries and relays.

Artificial systems such as technologies consist of objects – the relational
materiality of everyday life. Objects like the swimming pool robot cleaner
involve a form of social delegation and substitution. The cleaner is the
'designed' outcome of a trade-off between delegating to human cleaners
and non-humans. In Latour's approach, the robot cleaner then prescribes
what sort of people use the pool. The cleaner is part of one of many chains
containing intermediaries between humans and non-humans.

Networks require intermediaries such as texts, technologies and disciplined bodies that are *circulating* forms: immobile mobiles. Texts occupy an important role as intermediaries. Examples include maps, journal articles and pre-nuptial contracts. One example is the code used in computing because its menus and routines have been delegated and inscribed for human labour. Boundary objects are devices that establish stabilization in the network through interlocks and truces. Boundary objects co-ordinate different hybrid networks and enable co-working. These intermediaries are also sent out by the designers and suppliers to maintain order and form the network. Network builders have to persuade, force, invite and seduce actors and actants to take their allotted roles. The focus on network designers/builders suggests that they are active and the users are a passive group who need to be told (e.g. Corke 1985).

The analysis of heterogeneous networks should acknowledge the possibility of contest and conflict over their introduction into particular settings. The stabilization of the network involves many actors including non-human actors (Bijker et al. 1987). For example, the active role of users in shaping an organizational innovation is often neglected. Moreover, there will be many aborted networks. Even when established they are subject to collapse. For example, the sailing ship network after the introduction of steam ships in the 1870s. Heterogeneous networks are the outcome of struggles between unequal partners. For example, specific groups may oppose particular organizational innovations. Therefore we cannot assume that all the key actors are involved only at the design stage (cf. Bijker et al. 1987). There are intervening processes additional to design and use. For example, translation and commissioning (Whipp and Clark 1986: 90–91). Therefore we should reject symmetrical explanations that omit politics and power. Explaining the agency of a heterogeneous network is very difficult, especially in the case of the role of the users of an organizational innovation. This means that the framework of morphogensis examined in Chapter 4 needs to be inserted into the explanatory framework.

Technology as organizational process

This section presents three vignettes on the theme of technology as organizational process.

Missing modules in SCITS (MRP2)

Strategic Co-ordination Information Technology Systems (SCITS) seems to represent systems that are tightly articulated. Certainly the suppliers' discourse and rhetorics provide that impression. Their presentations, as in Figure 8.2, exude system tightness and integration. The suppliers, however, draw their revenue from selling modules and many of those modules are

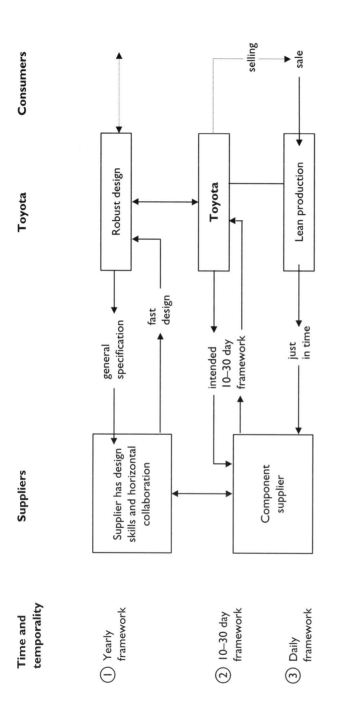

FIGURE 8.2   Three time-temporality layers at Toyota

specific to one area of the distributed knowledge of a firm. According to Buckley (1967) and Weick (2001), organizational systems are loosely coupled systems and it is organizational processes and mechanisms that activate and articulate their co-ordination. Field studies and ethnography of the firm show that both strategic and tactical co-ordination has been and is executed through organizational processes rather than the software. The point is underlined in two pieces of research. Burcher (1991), in a large-scale survey of over 200 plants, demonstrated that for MRP2 key integrating modules were typically missing or were not used. So how was co-ordination possible? Scribner and colleagues examined organizational co-ordination in laboratory studies and field studies. They demonstrated that different occupational roles existed in firms to handle explicit, structured and tacit, loosely structured knowledge. Some roles dealt in explicit knowledge and abstract models of the firm. Other roles were grounded in the language, power relations and normative frameworks that interfaced with specific departments. The interface between these different types of role and their articulation shaped the quality of co-ordination in the firm. Put simply, the organizational processes were as integral to SCITS as were the software packages.

Enterprise Resource Planning and UK universities

Universities have long been characterized as possessing social technologies that are primarily pooled interdependencies between departments rather than sequential or reciprocal (J.D. Thompson 1967). The observation is useful, yet is changing and an important ingredient is the new infrastructural material technologies of SCITS. American universities have perhaps resembled professional bureaucracies (Mintzberg 1978) rather than the possessive individualism of the British collegial system. UK universities have been less formalized, allegedly more parochial, and with strong vested interests. UK universities are moving to more corporate forms and the new managerialism. In the UK, the Joint Information Services Committee (1996) reviewed the overall position across all universities with respect to their information sharing and concluded that a new model was required. Their report coincided with the salience in the market place of SAP, the German supplier of systems with an accounting tilt. Moreover, most universities are advised by the global accounting firms and have adopted systems that provide an over layer of financial information. The new managerialism and positive political economy of the market enforced since 1981 stimulated that over layer, and transformed it into a system of making the teaching and research performance of each university more transparent.

Cornford (1999) examined the initial introduction of these systems into UK universities using the PIEM approach and heterogeneous engineering. They conclude that:

(1) The suppliers and the pro-innovation lobby constructed a contrast between the 'old model' and the 'new model'. The old model had reached its technical limits, it was unco-ordinated, unregulated, slow and old-fashioned, it lacked bridges between the centre and all the departments, it created chaos. The new model addressed the 'life blood' of information flows to create a complex model of the university through which, by adopting the new model, the university should transform itself. The rhetoric of the virtual university was raised. This served to problematize the existing system, especially in the financial and administrative centre of the universities.

(2) Project groups that incorporated members of user departments undertook the rollout. However, there was only a slight attempt at heterogeneous engineering.

(3) The partial adoption of the new model did make the detail and variety of local practices visible. The centre introduced more coercive forms of governmentality with a tightening-up of roles and procedures. Also, there was a strong pressure to standardization and commensurability of practice across departments. There was a failure to enrol a wider constituency. At the department level there was extensive 'working around' the official code inscribed in the software (Pollock 1998). However, there was no evidence of increased flexibility.

(4) A new hybrid of the formal infrastructure began to emerge.

Unfortunately, the analysis did not adopt the three-step realist framework (see Chapter 4) and so the interplay between pre-existing conditions and the social interaction was conflated. Even so, the emergence of a hybrid is interesting and very relevant.

Intranets

According to academic analysts and to some of its designers the Intranet is a web technology that creates a platform on which distributed knowledge held by functionally discrete groups can be both protected (the firewall) and also transported between these distributed groups. The platforms connect discrete groups and their explicit knowledge bases to the rest of the firm and to recognized centres around the world. The suppliers portray the Intranet and their enrolled network as de-centred with loosely coupled layers on several levels. The Intranet facilitates the spatio-temporal flows and combines distributed knowledge. This is undertaken within an immensely fast flow that retains the designated form of the immutable mobiles over wide spaces. According to this view, the Intranet is an altering innovation rather than an entrenching innovation (Clark and Staunton 1989). However, this is an essentialist perspective on technology centred on the efficiency hypothesis.

The Intranet has been rapidly diffused, but is actually located in contexts with a pre-existing political system and power structure grounded in distinctive systems of meanings. There are competing as well as co-operating groups. These relevant groups may have quite different cultures and strategic intents. The users, who may play a central active role, can unpack the Intranet and inscribe their own meanings on its usage. This might mean that they attempt to use the potential of the firewall to ring-fence their political position and reduce exposure to transparency. So, the incoming innovation is surrounded by multiple stories about its role. It is possible that the altering notion will be transformed into an entrenching situation. Newell, Scarbrough and Swan (2001) provide evidence of this happening.

Strategic Co-ordination Information Technology Systems

(1850s–2000)

The contemporary organizational innovation known as Enterprise Resource Planning (ERP) is the recent shape and form of strategic co-ordination information technology systems (SCITS) that were partly envisioned in mid-nineteenth century America (Beniger 1986). These are genres of control through communication (Yates 1989). This section provides a rough guide rather than a full, analytically-structured narrative. Periodizing the evolution from the manual proto-systemic practices right through to the full-blown SCITS covers *five eras*: pre-1850s; 1850s–1940s; 1950s–1970s; 1980s–1990s; 1995 onwards. For much of the narrative the USA is the key milieu for constitution of the innovation. Japanese and German (e.g. SAP) contributions become more evident in the past two decades.

Pre-1850s

After 1790 the American state continuously played a role in enabling the commodification of knowledge through legalism and in supporting contracts in the face of extra-legal claims (de Soto 2000). The state created institutions that pooled the pre-existing capacities for how-to-do-it recipe knowledge and reflexively constructed pragmatic knowledge. The state's role in the long-term capacity to produce interchangeable modules and components has been quite remarkable. These underpinned the emergence of the American System of Manufacture after 1830 whereby end-products could be assembled from standardized components with a design-led framework. America became the home of functional design (Pulos 1983). Patents and the ownership of intellectual property rights interfaced well with the capacity to insert models into texts and for the legal profession to protect the owners. Moreover, government establishments for defence-related material were centres and obligatory passage points for the contracting between the state and the private sector. Minutely detailed and explicit instructions were constructed and applied to suppliers at, for example, Springfield Armory. Also, the Connecticut Valley became the incubator for the development and subsequent diffusion. The social capital resonated with solutions to the extra-legal mystery of capitalism (de Soto 2000). By the time of the 1851 Exhibition in London, the basic principles of an American system of organizing were in place (Clark 1987, 2000). This is the proto-SCITS era. Prior to the mid-nineteenth century, American social capital about work organization and design discourse was quite exceptional when compared with Europe.

1850s–1940s

From the 1850s into the 1940s the genre of SCITS continued to develop structured methodologies and mechanical calculators (e.g. Gantt chart) increasingly supported these immutable mobiles. The growth of the administrative supply industry was a major feature in both supplying and transferring innovations between sectors of the economy in the search for profits (Yates 1989). In the late 1940s electro-mechanical machines were widely incorporated.

Returning to 1851, three British visitors recognized that there was an unknown system (a new genre) that they soon referred to as the American System of Manufacture (Rosenberg 1969; Hounshell 1984). By the 1860s professional engineers were developing the administrative sciences (Clark 1987). The role of the leading engineering professions in the building of the American state from West Point diffused systematic recording and reflection as a form of analytic thinking. In addition, the American proclivity for popularizing the Darwinian vision of Herbert Spencer centred on the commercial roles of systematic thinking and of knowledge creation. Structured methodologies were both open to innovatory problem-solving and were techniques through which solutions could chase problems. System management (1870s) spawned a variety of movements, including Scientific Management. The latter involved extensive manual charts for observing and critically analysing existing practices. Observation revolved around recording space and time. The latter being done exclusively with specially designed portable clock-watches. Photographing and then filming work systems, as well as sports, became routine. Systems for planning, such as the Gantt chart, became commercially available. Also there were many didactic texts. By the 1920s there was a whole series of heuristic rules. For example, in retail, all the shelf space was measured and then the financial flows through the space were recorded as total revenues in a given period minus costs. In this way the rates of return for space were located and stocking moved to those items which produced the most profit. In factories, the typewriter and calculator were extensively used in conjunction with specialized charts and tables to analyse material and financial flows across a wide range of areas.

Corporate timetabling became routine with regular weekly updates done at a hierarchy of levels within the specialized, separated departments of the growing bureaucratic organizations. These timetables sought to match expected orders to production requirements and used prototypes of Master Production Schedules. Chandler (1962) shows how the introduction of the divisionalized form of corporate organization enabled strategic co-ordination to focus risk and to search for heuristics and patterns that coupled the firm to its context in a particular form of administrative expertise (Penrose 1959).

1950s–1970s

Existing facilities have been constantly rationalized since the 1870s (Chandler 1977). Planning systems were well established by that time and major firms began to shift from centralized, single-product firms into the multi-divisional firm in the 1920s. This development created the reverse salient for division-wide planning and co-ordination. By the 1950s there was a large population of big firms attempting to manipulate large data sets. These offered reverse salients to be formulated and resolved. During the period electro-mechanical systems were incorporated for calculating the schedules and the overall financial state of the firm. Cost accounting was incorporated. In manufacturing, there was a well-developed heuristic knowledge about economic batch size and the economies of scale. During the 1950s there were major developments in administrative theory, with the founding of key journals, such as the *Administrative Science Quarterly*, and key academic institutions, such as the Institute of Management Sciences (IMS). These provided an infrastructure capable of relaying abstract templates and they extended knowledge commodification in the expanding supply side by providing templates for organizational professionals. These templates were additional to those associated with the highly prestigious engineering professions and with the growing sector of business schools. From the 1960s onwards computers and software began to play their central role. There was a pivotal shift from Economic Order Quantity (EOQ) to the steering heuristic of the Bill of Materials. The establishment of a software module for the Bill of Materials enabled Materials Requirements Planning (MRP), a production-oriented system which periodically (e.g. monthly) reviewed existing cost and performance against a timetable. MRP positioned sales and finance in the same framework as production.

The Cold War stimulated state involvement in the practice and theory of co-ordination through process technologies (Clark and Newell 1993). IBM and other firms developed the mainframe computer for corporate use. A related development added another trajectory to those trajectories already shaped by the state.

Organizational professionals from production planning, inventory control and related areas formed associations at local levels (e.g. New

Bedford, Massachusetts, 1946).[4] These merged in 1956–57 into the American Production Inventory and Control Society (APICS) when 26 professionals, most of whom already represented local associations, met in Cleveland and decided to create a national society. The first national APICS conference was held in 1957. APICS became an obligatory passage point for the new professionals. At that moment the membership comprised primarily practitioners in manufacturing. The network thereforee played a 'brokerage' role at the formal level in Rogers' two-level framework for diffusion and crossed structural holes with weak ties. APICS, through its journals, educational programme and local meetings, enabled knowledge of practice to be calibrated locally as well as local practices to appear in nationally read literatures.

During this period a handful of entrepreneurial practitioners working in a small number of large companies introduced key organizational innovations by exploding production requirements down through a Bill of Materials rather than controlling production through the use of EOQs. These firms and key intrapreneurs were developing in-house computer systems around the Bill of Materials into a new system that soon became known as Materials Requirement Planning (MRP): Burlinghame at Twin Disc, Alban at Black & Decker, Rosa and Kelly at Perkins Engines (Wight 1984). Although these individuals seemed to be isolated agents, they were connected through professional and personal networks to two IBM consultants – Joe Orlicky, who was manager of Manufacturing Industry Education, and Oliver Wight, a senior industry analyst. Each used his industrial experience before joining IBM to develop ideas about production planning and inventory control in a manufacturing rather than computing context.

Materials Requirements Planning was a simple programme which established a match between the product demand based on sales orders and the planned quantities of components required. The aim was to balance depletions in the inventory. MRP relied upon the Master Production Schedule (MPS) that was based on forecasted orders, but forecasts are

---

4 This section draws from the current research by Sue Newell (Royal Holloway), Jacky Swan (Warwick), Harry Scarbrough (Leicester) and Maxine Robertson (Warwick). It is funded by both the Engineering Physical Sciences Research Council (EPSRC) and the Economic and Social Research Council (ESRC). There is a useful website with Jacky Swan at the Warwick University Business School. I am especially indebted to them and to the IKON group. The initial studies, also funded by the EPSRC and the ESRC, were undertaken in my research groups at the Aston Business School and include David Bennett, Peter Burcher, Sudi Sharifi, Neil Staunton as well the (then) even younger Sue Newell and Jacky Swan. The research of this community also influences Chapter 9. The IKON grouping includes: Joe Tidd (Imperial College and SPRU), Nick Oliver (Cambridge), Fred Steward (Aston), Jamie Fleck (Edinburgh) and Steve Conway (Leicester).

rather uncertain. The numbers on which the plan was based were subject to inaccuracy and also needed expensive updating at intervals such as a month or a week. The system was open loop and did not take account of real time or actual use from the inventory.

In 1966 Joe Orlicky, Oliver Wight and George Plossl met serendipitously at an APICS national conference. All were active members, giving seminars on MRP and writing articles about MRP for the APICS journal. Quite independently of one another, however, each had decided to attend this particular conference specifically to disseminate and promote practice-based knowledge about MRP. George Plossl and Oliver Wight established at this initial meeting that, in fact, they already knew one another, having worked together several years earlier at the Stanley Works. However, prior to this fortuitous meeting all three men were not aware that they had all been working in very similar ways on the computerization of MRP. This highlights the highly informal, opportunistic character of this encounter so reminiscent of the early stages of innovation. It also shows how the development of strong ties (e.g. between Orlicky, Plossl and Wight) and weak ties (e.g. via membership of APICS) were conflated. This suggests that the distinctions between these different network forms may be somewhat more blurred in practice than the literature suggests. Orlicky, Plossl and Wight had common interests with regard to the operationalization of MRP by using bespoke computerized systems. They also shared concerns in that, while they believed MRP was a superior approach for Production Inventory Control, it was in fact a new paradigm that required a fundamentally different way of thinking about Production Inventory Control (PIC). Despite the successful implementation of this approach in companies they had worked for, and subsequently as consultants, they were acutely aware of the lack of knowledge about MRP in manufacturing at this time. The vast majority of large manufacturing companies in the USA had not begun to consider using MRP. Thus, despite the fact that to some extent they competed as consultants in the same business, these individuals had much to gain from pooling their expertise in MRP at this early stage of idea generation (Kreiner and Schultz 1993).

Orlicky, Wight and Plossl had each come to the APICS conference in 1966 in the hope of lifting the veil of ignorance around MRP by presenting the results of their work in a variety of manufacturing environments. Following this chance encounter the three decided to join forces on an informal basis to develop and promote the MRP design template. Initial variants were thus primarily designed at the local level of the manufacturing plant, characterized by high levels of interpersonal networking and weak ties, with some direct involvement of IBM. In addition, the professional network provided by APICS at this time played a crucial role in the development of networking relations among the 'group of three' who then became focused on further developing and diffusing knowledge around MRP. In this instance, then, the institutional context of the professionalization of work in PIC, coupled with the specific network structure of APICS, provided weak links among organizational actors. Interpersonal networking activities among the group of three and among the original founders of APICS began to generate more direct or strong ties among subsets of actors as clusters of firms began to enlist the help of Orlicky, Wight and Plossl in MRP projects. However, in this case it was the dynamic interplay between professional network structures and interpersonal networking activities that shaped this phase of the innovation process.

MRP was rapidly diffused to large American firms and then to specialized firms overseas. For example, a variant was used by Rolls Royce in the late 1960s and its application revealed the impending financial disaster of the RB211 engine design. This period closes with the commercial viability of MRP systems in the 1960s.

1980s–1990s

MRP evolved into Manufacturing Resource Planning (MRP2) in the 1980s. MRP2 achieved closure with a new module, Capacity Requirements Planning (CRP), which took into account shop-floor capacity when producing the Master Production Schedule. The closed-loop connectedness of MRP2 was aesthetically displayed to potential users by showing the connections between the modules as a metro-like collection of journeys. The display is intended to evoke a sense of control and practicality while omitting attention to the necessity for the organizational infrastructure to embed MRP2 in existing practices. The principles of MRP2, as with MRP, involved IBM in training members of APICS and encouraging them to 'spread the word' at branch meetings.

MRP and MRP2 were designed to provide the technostructure of the firm with a sense of having a high degree of control over events. This vision of control is an integral part of the American Dream (Beniger 1986; Nye 1997). The suppliers offered a vision of 'total information' in 'real time' coupled with the centralization of decision-making and surveillance of the shop-floor by the 'sovereign' centre. Many departments of the firm seemed to be controlled in the same plan: sales (top left), costs and revenue (mid-right), shop-floor (bottom right), purchasing (bottom left).

Four features of the use of MRP2 were understated:

- MRP2 required an immense infrastructure and discipline to maintain its many databases
- In many instances the full closed system was not used and sometimes-key modules were not even purchased (Burcher 1991)
- Because MRP2 was production-oriented, members of the marketing and the production professionals rarely enrolled accounting functions
- Making the system appear to be a fully closed loop and also usable required a blending of several occupational roles. Otherwise this was rhetorical closure (Bijker et al. 1987).

The feature of MRP/MRP2 that has been widely examined is its different usage in certain Japanese firms (Clark and Newell 1993). Japan was drawn into the American political economy after 1945 (Dower 1999), especially for a role as a supportive infrastructure in the Cold War after the Korean conflict. At that time Japanese manufacturers faced similar problems of heterogeneous markets to those faced by many British firms. Whereas British firms often sought to emulate American solutions, certain Japanese firms became the milieu in which the reverse salients of Fordist mass production were overturned. Key Japanese managers sought to develop forms of production suited to small batches. The solutions they developed over several decades both transformed existing notions of economies of scale and opened the window on mass customization. Key events occurred at the micro-level that enabled some firms to develop visions, infrastructures and techniques directed at the problems of small batches, scarce spaces, expensive inventory and a rather poor indigenous market containing unique demands on products. Imported British saloon cars performed poorly in mountainous, winding-road conditions analogous to Italy (Cusumano 1985). In addition to these covert and developing capabilities, Japan probably benefited from the cultural import of American management gurus and techniques, even though these were ill-matched to the local corporate infrastructures. The Japanese did not emulate; they unbundled American approaches and selectively appropriated key elements which were then inserted into local infrastructures (Clark 1987, 2000; Clark and Staunton 1989). These micro-level and meso-level developments were coincidentally supported by the active state policy of intervention at the macro-level (McMillan 1985).

The differences between the USA and Japan became especially evident in saloon car production. The Americans achieved customization by an expensive and lengthy process of adding in variations to satisfy customers and thereby disrupting the high-volume production system, especially at the final assembly stage (Abernathy et al. 1983). Japanese companies like Honda simplified this complexity by offering packages of options, thereby reducing the complex synchronization of the assembly lines and the feeder tributaries from the suppliers. As is well known, Japanese firms evolved forms of pull rather than push approaches to production. The pull systems relied on simple documents to activate the next stage and to create Just in Time (JIT), as well as reducing the time taken to market cars. Also, by using many parallel specialist groups within carefully articulated, metaphor-laden design teams, they reduced the design cycle time and increased the speed of producing new models.

How did MRP and MRP2 fit into the Japanese automobile firms? Toyota operated its production system in three major time frames: the year, the 10–30 day period, the day. These are shown in Figure 8.2. The yearly frame articulates around robust design (Clark and Staunton 1989) to share general specifications with the top tier of suppliers who possess design skills. MRP2 systems provide a useful infrastructure, especially in a context where the managerial division of labour seems to be more integrated and less conflictual. At the level of the month down to ten days and the week, MRP2 systems are able to provide a reflective account of events. As MRP2 replaced MRP, then Toyota moved down from the month to shorter periods of reasonably exact accounting. This allowed them to co-ordinate with their component suppliers on the near future. The third level of JIT was a superior functional alternative to the American desire for daily control. The Japanese possessed reasonably exact synchronization, sequencing and pacing from JIT and did not require the production and centralization of large amounts of accurate data. They both saved money and also used their expenditure more profitably. So, contrary to some accounts, the Japanese were able to benefit from MRP and MRP2. However, American learning and the further evolution in SCITS undermined their advantage. The software of MRP2 systems continued to evolve and there was a step change with the development of Enterprise Resource Systems (ERP) and their emergence was part of a qualitative shift.

1995 onwards

From the mid-1990s onwards the Internet, Enterprise Resource Planning (ERP) and surveillance systems exemplify one element in the new very visual, flexible and rapid possibilities in SCITS as the engine of mass customization. For the first time there is a certain European dynamic arising from the central role of SAP in commercializing ERP around the finance and accounting pivot rather than the production pivot. ERP was an extension of MRP2 systems but then widened the scope. ERP differs in its use of relational databases and object-oriented programming language as well as in engineering tools. ERP increases open system portability. ERP systems are accounting-oriented information systems that symbolically translate all action into money. ERP systems identify and plan the enterprise-wide resources needed to take, make, distribute and financially account for customer orders. SAP both leases these systems and franchises their distribution by enrolling from a sophisticated and extensive labour market of consultants. ERP should reduce overall costs and give an IT focus that is accessible and updated continuously. The aim is to provide a platform for the development of e-commerce and to include such options as video and bundled voice applications. It is these systems that are being cascaded into global public sector organizations.

Link

Chapters 7 and 8 conceptualized and explained the roles of two of the three players in the Decision Episode Framework. The third player, the user, is analysed in Chapter 9.

The Decision Episode Framework shown in Figure 9.1 is a configurational field containing many hybrid networks and is therefore a collection of loosely connected, slippery and autonomous actors. The two previous chapters have examined the pool of innovations and the influential gaze of the various suppliers. This chapter conceptualizes the user firm and also firms on the supply side. There are four sections.

First, the chapter commences by defining the firm and the market. The early polarization was between the rational model and the community model. The rational model has been adapted to embrace cognitive models and the learning organization. The community model can be aligned with both the relevant group perspective of Bijker and the community of practice model of Cook and Brown (1999). My definition of the firm will offer a synthesis weighted towards the community pole. In addition, there will be strong distinctions between 'what is', 'what ought to be' and 'what can be' (Clark 2000). Particular attention is given to the issue of organizational repertoires and their influence on the capacity of the firm to absorb innovation and to adapt in a co-evolutionary context. But repertoires cannot be constantly adapted. Most firms fail through exit or takeover. Selection-out is at least as prevalent as adaptation.

Second, we examine the distribution of expertise around the players in the configuration field. The left-to-right models of knowledge locate the centre of relevant expertise in the universities. Now the argument is that there are varieties of expertise and they are widely distributed. This contention is exemplified in the new production of knowledge (Gibbons et al. 1994). Consequently, if the universities are no longer the monopoly suppliers of expertise, then how are the various distributed actors connected on particular projects? The notion of knowledge-chaining illustrates certain features and highlights the search for new conceptualizations of knowledge chains (see Figure 9.2 on page 174).

Third, three examples of technology as process are presented to link into the next section in the form of vignettes.

Fourth, the Decision Episode Framework (DEF) builds on the community and knowledge-chaining model of the firm. DEF seeks to unpack the dynamics of the interaction between users and suppliers. Suppliers use various structured methodologies and theatres of consumption to seduce and enrol the users, but their configurational field contains high

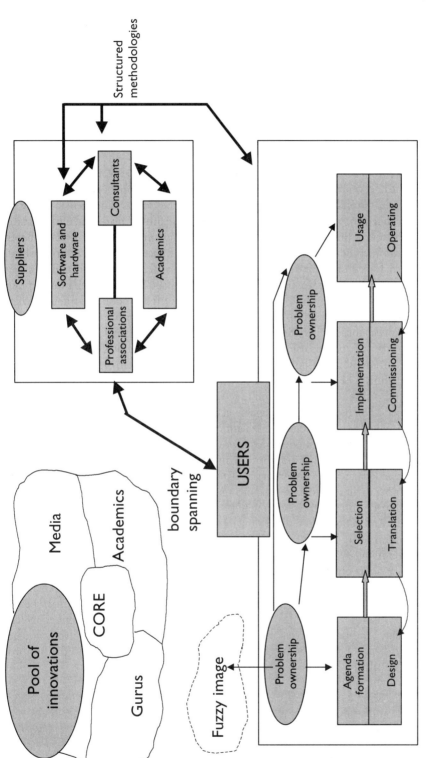

FIGURE 9.1 User–supplier dynamics (modified from: Clark et al. 1992–93)

uncertainty and equivocality. Exponents of the rational model issue didactic prescriptions yet these only rarely engage with the heterogeneous hybrid networks that have to be enrolled. This problem of process cannot be easily resolved.

Conceptualising the market and firms

Fligstein (1990) and Callon (1997) apply the perspective of the new political economy to the firm and the market. In their approach markets require political forces to operate. Markets are institutionalized systems of rules that determine who has claims upon the profits of the firm. The state is central in articulating the general rules of competition and market-specific definitions of the governance structures of firms. Contingent contracts represent one way of ordering the processes of conflict. The rules of exchange specify who can transact with whom and the conditions under which transactions should be carried out. There are economic actors with shared conventions about interpreting situations in order to inhibit actions that might undermine the market they have defined (see Chapter 6). These elements are necessary for markets to operate. Sociologists argue that markets are socially constructed and politically maintained. Therefore the state plays a central role in its own territorial domain even when that role depends upon the rules of exchange shared by a number of states (see Porter 1990). The transactions between states resolve the conflict by defining a price. There is a process of translation (equivalence) between conflicting interests. Consequently, decision outputs (prices) are highly distributed with clusterings around pivotal firms. The market is a cluster of regulated institutions and a co-ordination device in which agents pursue their own interests, being guided by economic calculations. There must be agents capable of calculation from which the output is the price. The agents as buyers and sellers have divergent interests. Callon (1997) observes that the notion of an economic transaction depends upon breaking away from local embeddedness. Those firms that are likely to succeed overseas will occupy a special position (Clark and Mueller 1996).

Based on our analysis of the organizational processes in Chapters 1–6, the firm may be defined in the following way (Clark 2000: 102–103). In this definition a distinction is drawn between what a firm is like according to theory laden diagnoses (the '*is*') and the ideals according to theory (the '*should be*'). There is also the issue of whether any gap between the 'is' and the 'should be' can be closed (the '*can be*').

(1) Firms are working communities containing mutually antagonistic and common interests that are connected to the struggle over the control of property rights inherent in capitalism.

(2) Their employees provide a good or service to a defined, yet varied population of widely varying stakeholders, including end-consumers.

(3)   The working community is embedded in a national cultural reper-
      toire and national innovation-design system. Each of these contains
      mechanisms affecting their re-production or transformation of exist-
      ing practices. It is likely that the national market is a major actor in
      shaping the expertise and learning of the firm's top management.

(4)   The working community is increasingly organized through abstract
      principles and templates of best practice contained in a discourse
      from consultants who claim to provide the more salient analysis of
      value-adding activity.

(5)   The firm is always in competition with alternatives yet is also in co-
      operation with firms in its network and chain.

(6)   Unless the working community can construct a robust constitution
      for resolving conflicts of interest, then the firm is likely to be
      selected-out from its competitive domain.

(7)   Firms specialize in speed and efficiency in the creation and transfer
      of knowledge. Firms can have more capabilities than markets in the
      assembling of tacit knowledge.

(8)   Firms are specialized assemblages for organizing and achieving an
      organizational advantage through structured co-ordination and
      communication about organizing (Nahapiet and Ghoshal 1998).

(9)   Firms are located in networks and chains along supply lines. Some
      firms may be pivotal in shaping parts of the chain and even shaping
      the chain as an overall structure. Ikea shapes the actions of a variety
      of suppliers across Europe.

(10) Firms are often multi-sites with sites in different nations. This
      extended network requires capacities to enrol and control the
      discretion of distant units while also adapting to local contexts and
      accumulating learning.

(11) Action is organized recursively through assembling recurrent action
      programmes from the organizational repertoire.

(12) The existing repertoire of capabilities frames the starting line for the
      capacity of a firm to absorb new capacities through upgrading and
      innovation.

This concept of the firm differs from that in organization theory and design
(e.g. Daft 1997).

   This community-oriented definition can also embrace the core features
of other approaches. The influential approach known as *communities of
practice* (as Brown, Cook, Duguid, Hutchins, Lave and Wenger) can be
situated in the '*relevant groups*' around the evolution of an innovation
(Bijker 1995). This means that analysis has to identify the relevant groups
in a configuration and their fuzzy, overlapping and competing boundaries.
For example, in English hospitals the problem of legionnaires' disease
embraces an array of relevant groups within the hospital (e.g. nurses,
doctors, administrators and the Trust Board), in Birmingham University
and at various state-owned centres for public health (Clark 2000: 255–
256). The definition requires an extension to include the *relational*

materiality of the social made durable: technology, geo-location, buildings and equipment.

Much of the community definition is readily translated by conceptualizing organizational action in terms of organizational recursiveness and repertoires (Clark 2000). Recursiveness appeared in the language of analysis in the late 1970s (Clark 1978; Giddens 1981). Since then recursiveness has been used very loosely to refer to elements of repetition in organizational processes. These are rarely exemplified. Recursiveness is a repertoire of organizational processes. All firms possess such a repertoire – just as sports teams do – but the repertoire varies in scope and richness (Clark and Staunton 1989: Ch. 9).

Absorptive Capacity and Antecedent Repertoire

Cohen and Levinthal (1990) conclude that firms are more likely to borrow than to invent, but that borrowing depends upon the capacities they already possess. The absorptive capacity of the firm is strongly shaped by the antecedent repertoire of the firm. The firm 'needs prior related knowledge to assimilate the new knowledge' (ibid. 1990: 129). The capacities in the repertoire will be those that are distributed throughout the firm and are capable of being articulated. The notion of requisite absorptive capacity refers to a future scenario and the issue is whether the members of the firm can acquire those capabilities before the onset of their being required. The antecedent repertoire is domain-specific and the outcome of a path-dependent, historical and cumulative process. Absorptive capacity mediates between spending on Research and Development and the interdependencies with competitors (e.g. spillovers, extra-industry knowledge). Two comments are relevant. First, this useful concept is tilted towards downward conflation into the past and is weak on agency and emergence. Second, the concept should be situated in the interactive chain-link model of knowledge (Kline and Rosenberg 1986).

Selection/adaptation and frictionless change

Most theories of innovation and change imply that there are simple, powerful models whose application is achievable provided a few basic principles are applied. The best-known models are in the post-Lewin school (Hatch 1997). These tend to focus at the organization level and to give light attention to history and context. Studies and theories of innovation are pervaded by teleological theories of smooth change and adaptation. Countering these claims has been the province of population ecology theories (Ziegler 1996), the selection/adaptation issue and soft-determinism.

The selection/adaptation issue raises three main problems for a useful theory of innovation. First, there is extensive longitudinal research to show that the population of firms in any given sector typically experiences the entry and exit of firms. Typically, more than 90 per cent of new firms founded two decades ago have exited or been taken over. Very few of the survivors grow beyond medium size. Moreover, even long-established firms do exit. Second, the long-wave theories of innovation all make the Schumpeterian assumption that the capabilities of existing organizations may be challenged so strongly by the periodic bouts of epochal innovation that there is 'creative destruction'. In effect, pre-existing organizations either disappear or experience dramatic transformations in their knowledge bases, structures and processes. Third, marketing surveys of the launch of new products suggest that at least one-third do fail.

The conclusion has to be that adaptation by firms through innovation is a highly risky process. It seems that many firms are selected-out from their sectors (Aldrich 1979). Even so, case studies are typically of successful firms and product/service launches. We require a useful theory of innovation that incorporates the adaptation/selection issue.

Designed performative knowledges

New Production of Knowledge

In his essay on new forms of knowledge, Lyotard (1984) introduced the notion of performative knowledge and sketched a future for academic capitalism in which the locus of performative knowledge became anchored outside the universities. A similar theme is explored in *The New Production of Knowledge* (NPK) by (Gibbons et al. 1994; cf. Huff 2000). The massification of education has created immense capacities for reflexivity in non-academic settings. The university's monopoly of knowledge production is over because there is a parallel expansion of the number of knowledge producers (Gibbons et al. 1994: 13). Consequently, new forms of knowledge are produced outside the universities in '*contexts of application*'. Now there is competition over the production of performative knowledge.

Gibbons and his colleagues contend that the new mode of knowledge production is different and is diverging from the old. Knowledge is increasingly shaped by its origin in contexts of application. The role of contexts of application relative to the universities is very significant. For instance, economies of scale are displaced by economies of scope.

The universities are professional bureaucracies using economies of scale to distribute *segmented* knowledge. University knowledge is:

- Identical to the enlightenment concept
- Produced in special centres away from contexts of application
- Created by disciplinary-bound, bureaucratic professionals

- Applies cognitive norms and practices to determine what constitutes good scientific activity
- Driven by the scientific definition of performative knowledge.

This knowledge is representative of and architecturally part of the economies of scale. Many universities still do relatively little to develop distribution and have become isolated from all except their student populations. The diffusion patterns of universities rely on being drawn into benefits without being required to target their knowledge production. Consequently, the interaction of research with production-consumption was sustained by heavy parallel investments in production and distribution. Critics of this mode of knowledge distribution claim that there are huge chunks of segmented knowledge that are largely inapplicable. Moreover, universities lack the full understanding of the context of application and of networks of action within which such knowledge could be made actionable. Therefore we should recognize the constraints of production in universities and look to the potentials of the alternative, which is produced outside in contexts of application.

In NPK knowledge is organized around particular applications and is therefore embedded in a context of application (e.g. pharmaceuticals). It has become designed knowledge: tightly integrated rather than segmented. *Designed knowledge* is intended to be:

(1) Useful and accountable to the many stakeholders
   - produced under a process of continuous negotiation
   - in a complex demand–supply relationship, yet beyond the market.
(2) A fusion of intertwined, complex transdisciplinary knowledges.
(3) Articulated through *context-specific frameworks* that are used to guide problems. The solution might arise outside the context of application.
(4) In the form of solutions that are equivalent to knowledge though not necessarily disciplinary knowledge because NPK is:
   - Dynamic and diverse,
   - Heterogeneous and subject to reformation,
   - Highly reflexive,
   - Located in temporary teams that are constantly being formed and dissolved.
(5) Agglomerated in communities of practitioners. Consequently the 'findings' are more in the form of heuristics than in the form of tightly related propositions favoured by the academic journals. NPK is anchored in the heterogeneous ontologies and is communicated through conversations and practice. Quality control is therefore undertaken in the context of application.
(6) This process requires a deepening and expansion of scope into depthful ontologies in order to produce actionable knowledge.
(7) Quality control is produced in the context.

(8) Knowledge is cumulated through 'the *repeated configuration of human resources in flexible, essentially transient forms of organization*' (Gibbons et al. 1994: 9, 14).

Gibbons et al. (1994: 9 and 14) emphasize that the New Production of Knowledge is cumulated through the 'repeated configuration of human resources in flexible, essentially transient forms of organization'. These features are exemplified in the growth of the design approach to drug production in the past two decades. The emergence of Mode 2 reflects the growth of academic capitalism and creates challenges that require existing national institutions to be de-centred from the universities into firms and the wider society (Nowotny et al. 2001).

In firms, their puzzle-solving regimes are directed towards admixing and fusing explicit and tacit modes in a shift towards modes of enquiry oriented towards contextualized results. Consequently, it is the strategic principles of design developed in the context of application that guide the experimental process. The firm's knowledge includes both the expertise of the workforce and the focused knowledge used in the puzzle-solving regimes to create the core transformation processes that the firm is exploiting. This tends to obscure the central role of knowledge in the firm (Clark 2000). The centre of interest here is how the firm's strategic agenda defines a *specific design configuration*. The specific elements of puzzle-solving in the design configuration include both proprietary knowledge protected by patents and tacit knowledge implicit in the culture of the firm. The tacit knowledge will influence how proprietary knowledge is interpreted. It is important to refer to the dynamics within which components are frequently re-arranged as part of incremental innovation. Because NPK is carried on at the frontiers of application, there is less attention to first principles and more attention to '*ordered structures*' and configurations. The design-led search for architectures implies new forms of knowledge, especially through computational modelling.

In the NPK situation for *economies of scope* there are major and dramatic differences. In economies of scope the gains arise from repeatedly configuring similar and different technologies and skills to satisfy different market demands. Firms here are in networks whose sustained membership requires access to many arenas through a dynamic and also an evolving set of specialized capabilities. Thus the differences between distribution and production are broken down so much that customers are also involved in being incentivized to be part of the economies of scope. In this situation universities are exposed to producing non-usable knowledge for which they eventually are not paid. This arises because of the rising costs for research and the necessity for the public investment to be audited and regulated. Universities will have to develop new capabilities and to understand the competition over knowledge delivery.

Competition does play a role in the race for actionable knowledge. The analogy of the race involves three elements:

- A set of rules to determine legitimate behaviour
- The behaviours of competitors
- Criteria defining success and failure.

In the economies of scope the three elements evolve in an unpredictable manner, yet the regimes of knowledge production in NPK lead the firms to search for means of capitalizing their knowledge base. A dominant competence is the specialized, highly structured and specific knowledge platform and the capacity of its architectural configuration to enable the enclosed modules to be reconfigured for each new customer project. Firms and their networks must specialize and the creativity of the knowledge base is of dramatic relevance. Platforms are design configurations and form the heart-and-brain around which the firm unfolds its revenue-generating activities. The choice of a platform of knowledge is partly unintentional and yet is strategically consequential. Once established – rather than chosen – the configuration of knowledge in the platform locks the firm into a finite set of offerings to its customers (Henderson and Clark 1990).

NPK reveals how far the knowledges of scope are shifting the locus of survival from the firm into an inter-organizational network: the 'business-to-business' (B2B) perspective. For example, the nature of reputation and brand image (Schultz et al. 2000) in the situation of economies of scope is quite distinctive and is also under-recognized. Knowledge fora occupy a special place as arenas in which specialized scope knowledges are represented and commercialized.

The augmentation in the universities by NPK places greater emphasis on the fragmentation of knowledge into a spectrum of abstractions, experiences, data, dynamic figurative simulations, practices, theory and emotional elements (Clark et al. 2000). There are few clear rules, boundaries or settled routines. There is much fusion and hybridization. The former isolation of the cognitive and tacit modalities no longer applies. The spectrum of knowledge is explacit (Clark 2000). Consequently, new forms of collegiality and managerialism emerge to handle the new fusions exemplified by mechatonics and similar developments. This is reflected in the ways in which knowledge management increasingly moves on from data warehouses into the translation of the firm into its symbolic facets.

The capability to undertake NPK has already surfaced and its exponents are changing the range of transfer mechanisms to include:

- University patent offices being active
- New approaches to leveraging intellectual property rights such as equity ownership
- Liaison programmes
- Commercial sponsorship of research groups
- Round tables of diverse commercial interests
- The involvement of universities in regional planning.

Firms are increasingly examining their capacities to absorb externally generated NPK types of knowledge through co-operative alliances in the division of knowledge and through new power relationships in outsourcing. To some degree these new networks resemble football teams in which certain universities are players. However, many universities are still on the benches or in the bleachers.

NPK occupies a central role in addressing the risks of corporate investment in research. Firms have reversed the earlier tendency towards vertically integrated and managerially controlled firms. Firms now employ problem identifiers, symbolic analysts and knowledge brokers to commercialize their investments (Demarest 1997). There are data, information, learning and knowledge (DILK) at the core of the firm. Knowledge is traded and there is a division of knowledge based on continuous innovation.

The most important decision of the firm is how its choice of partners influences its 'design configuration'. The firm needs to design and create specialist knowledge in an explicit form. Because there is a distribution of knowledge, there is a potential problem of how knowledge moves around. Also, what are the dynamics of knowledge creation and consumption? Now management has to be for innovation rather than the source of innovation. Management has to resolve the issue of the intelligent consumer/customer and the network as the consumer. The role of the network alters the traditional user–supplier relationship. Designed knowledge requires new forms of integration and differentiation.

Hypertext organizations: Nonaka and Takeuchi

Two elements of the knowledge-creating approach of Nonaka and Takeuchi (1995) are relevant. First, their suggestion that firms should regard managerial work as routinely including the making of knowledge of different kinds. This idea introduces a designed element into Penrosian learning (Clark 2000). Their suggestion does localize the responsibility yet falls short of assembling the relevant mechanisms and practices. The mechanism can be located at the interface of the four states of knowledge creation (see Figure 9.2). The four states relate:

- Tacit to tacit
- Tacit to explicit
- Explicit to explicit
- Explicit to tacit.

In their own approach, Nonaka and Takeuchi seem to have knowledge and knowing effortlessly being created from a 'spiral' round the four states. The translation process whereby knowledge creation is achieved remains obscure (Clark et al. 2000).

Second, Nonaka and Takeuchi propose a new assemblage of the hypertext organization (1995: Figure 6.3). This combines three different

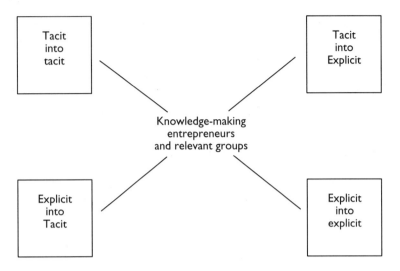

FIGURE 9.2   Explacit spectrums of knowledge and knowing

modalities of knowledge and knowing in the firm in a consciously designed action intended to reproduce some of knowledge creation strengths associated by them with American Football. Their use of American Football as a guiding analogy is inspired (see Clark 1987), contradicts their criticisms of western approaches and is awkward. American Football possesses both a game plan used by the head coach to inform the brief percentage of time when the club is actually engaged in direct confrontation with another team. The remainder of the time the club is practising, rehearsing and engaging in a reflexive monitoring of its many databases (e.g. videos, imagineering). The specialist coaches are seen as a key site for upgrading performance and for developing innovative plays from the generative expertise of the club. The hypertext organization contains three interconnected layers:

- Knowledge-base layer with databases, technology, culture and vision
- Business-system layer 'playing the game'
- Project layer containing project teams promoting knowledge creation linked by collaboration.

The project layer should engage in the continuous creation of explicit knowledge from tacit/tacit (see Figure 9.3). They form a 'hyper network' across the layer of the business systems. Individuals will have a high access to the knowledge base and its stored knowledge, but it is the project teams who trigger dynamic knowledge cycles and relate that knowledge to the corporate vision.

These two elements can be undertaken through the Intranet and through process technologies like Lotus Notes. However, the organizational process is not free from the features central to our earlier definition of the firm.

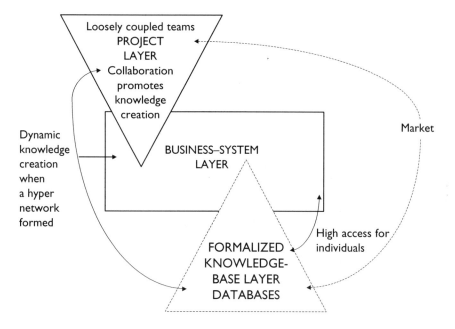

FIGURE 9.3 Hypertext organization (influenced by: Nonaka and Takeuchi 1995, Figure 6.3)

Interactive chain-link models

This section addresses the Kline-Rosenberg (1986) six-arena framework and shows how it explains the processes of innovation and design. This is a multi-state process of radical and incremental innovation of outputs, processes and work organization that involves differentiation into specialized units and also distinctive modes of co-ordination. The framework integrates:

- The past–present into the present–future
- Activities undertaken in organizationally divergent arenas.

There are a small number of studies examining the longitudinal dynamics of the innovation-design process in firms, but not in whole networks.

The innovation-design phase of new products and services requires attention to the connections and integration between the differentiated activities within networks and major firms. The issue of differentiation and integration can be examined through the Kline and Rosenberg framework that emphasizes the interactive character of knowledge-making in situations where knowledge is distributed. Kline and Rosenberg devised the framework to explain incremental innovation, but the same model can also be used as a framework for more radical innovation.

175

The model of the firm conceptualizes innovation design in six distributed arenas, each containing distinctive forms of knowledge. Figure 9.4 shows basic research from advanced centres in the upper box and another five arenas in the lower box of the firm. The model is shown as chain-links to express the interactive flows of knowing, knowledge and information that are likely to create economically valuable knowledge:

- The central chain of innovation-design is shown by the letter 'C' in the lower box, running from potential market on the left through to distribution and marketing in the right-hand box.
- There are various feedback loops. The most important feedback loop is shown with a capital 'F' running from distribution back to market potential. Secondary linkages are shown with small 'f' and six of these are indicated in the lower third of the model.
- The single most direct linkage between research to invention and analytic design is shown by the letter 'D'. The presumption is that this is shaped by the previous state of examining the potential market. However, that degree of neo-rationality probably underplays the iterative ways in which potential market is assessed and re-assessed.
- The 'I' and 'S' on the right-hand side, link distribution and market back to research. The 'I' refers to the provision of instruments, machines and technological procedures to scientific research. The 'S' refers to the support of research in sciences underlying the product (service) arenas to gain background information directly by monitoring outside work.
- The 'K–R' linkages from the knowledge domain in the centre of the model show the links through knowledge to research and the return pathways. So, if the problem is articulated and addressed between nodes '1', '2' and '3' and resolved at node '3', then no research is activated and the lines are solid. However, if the problem is not resolved at '3', then the research linkage is problematic and is a dotted return path shown as '4'.

The chain-link model does open up some of probable interactions and fits well with the claim by Cohen and Levinthal (1990) that R&D not only generates innovations but also develops the firm's ability to assimilate and exploit knowledge that has been generated externally: the absorptive capacity.

There is a latent tendency in the chain-link model not just to separate different areas in the spectrum of knowledge, but to imply a priority in problem-solving. This is useful in showing some of the contributions of Research and Development but underplays the division and spectrum of knowledges in a design-driven world.

Hage and Hollingsworth (2000) introduce a three-fold extension to the Kline–Rosenberg six arenas model to include more radical innovation in networks of firms where each firm tends to be much more specialized in its role for the network than was the case only a decade ago. They revise the six arenas slightly and interface the meso-level of network into the macro-

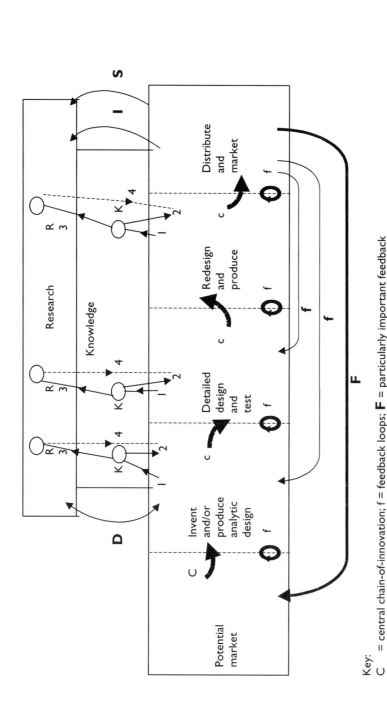

Key:

C = central chain-of-innovation; f = feedback loops; **F** = particularly important feedback

K–R = Links though knowledge to research and return paths. If problem is solved at node K, link 3 to R not activated.
Return from research (link 4) is problematic, therefore dashed line

D = Direct link to and from research from problems in invention and design

I = Support of scientific research by instruments, machines, tools and procedures of technology

S = Support of research in sciences underlying product area to gain information directly and by monitoring outside work

FIGURE 9.4   The 'chain-linked' model showing flow paths of information and co-operation (slightly modified from: Kline and Rosenberg 1986)

national level of the international economy. In the case of radical innovation, there is a deep division of knowledge in the network of many firms in each of the six arenas shown in Figure 9.4. The arenas are highly differentiated from each other. In the network level the actuality of different firms poses considerable potential problems. There is a likelihood that radical innovations will prove more awkward to accomplish because of their impact on the many different parties. Hage and Hollingsworth examine the shape of the radical innovation network. This suggests that the higher the number of highly trained researchers and the levels of expenditure, then the greater the capacity of the network to absorb externally derived new knowledge. However, this recipe is influenced by the connectedness between the six highly differentiated arenas. Also the specific character of national institutions that embed the six arenas will probably impact the connectedness. For example, the mode of co-ordination in South Korea is more likely to be through corporate hierarchies and in Taiwan through state-corporate hierarchies. This dimension can become very complicated when the six arenas are spread across more than one nation.

Network builder and network user

The Decision Episode Framework (DEF) shown in Figure 9.1 complements, augments and partially reformulates Rogers' (1962/1995) approach (discussed in Chapter 7). Before examining the framework it is useful to summarize some of the main contrasts between successful and unsuccessful innovation and then to lead into the DEF.

Appropriation processes

First, there is a distinction between successful and unsuccessful organizational innovation. Unsuccessful innovation is associated with a mindset that believes in innovation as an objective technical solution designed by experts, especially the system suppliers and specialist advisers. The unsuccessful project teams are likely to promise a great deal. They are likely to contrast old-fashioned systems unfavourably with the essential features of the new future system: the pro-innovation bias. So, knowledge is treated as codified, politically neutral and easy to transfer through structured methodologies. Although there may be some attempt to enrol support, the approach is unlikely to be influenced by the heterogeneous engineering concepts examined earlier. Also, the external suppliers and internal project teams tend to be given a lot of power and are likely to suppress conflict.

Successful innovation requires the users to adopt a discriminating view of the supplier's offerings and the supplier's rhetoric of pro-innovation. Appropriating new systems means contextualizing the innovation (Clark

1987: 153f.). The approach needs to recognize that knowledge is distributed in relevant groups which, to varying degrees, own problem-solutions and occupy positions in a stratified political reality. Because technology is a *process of organizational innovation* the processes need to be understood as interactive and political as well as cognitive. This involves knowledge creation, possibly through turning tacit knowledge into explicit knowledge (Nonaka and Takeuchi 1995). Successful approaches are likely to recognize the three-stage possibilities of the interplay between the pre-existing and the agency of social interaction leading to emergence (as Chapter 4). This means being astute in recognizing the emergence of useful hybrids. Successful innovation is likely to involve reflexivity about what is retained as learning from the experience of technology as process. Castells (1996) suggests that this politically nested process can be conceptualized as articulation and articulation needs to be conceptualized separately from appropriation. Articulation involves mobilizing social networks across distributed departments to address the distributed interfaces of knowledge by developing mechanisms. The PIEM perspective suggests processes covering meanings, legitimization and power.

Appropriation is a central process through which the groups and individuals from the user firm engage with the suppliers in translating the suppliers' version of the innovation. Suppliers and users are rather more active than is implied by the diffusion of innovation perspective (see Chapter 7). Suppliers engage in learning from users (Von Hippel 1988). Users increasingly attempt to develop their absorptive capacity by building internal and external networks. Organizational innovation involves many interfaces between distributed expertise. These create different understandings of what the innovation can do. Different groups will enact the situation differently (Weick 2001).

The problem of appropriation has been simplified in three major approaches. First, Zaltman and Duncan (1977: 268) elegantly formulated the clear separation of diffusion from implementation in their contingency model for designing organizations for innovation. They contended that the diffusion-selection phase is highly uncertain and should be managed with organic systems while the implementation stage is less uncertain and can be managed through mechanistic systems. The sharp separation of selection from implementation was reformulated by Eveland (1979) and incorporated into Rogers (1962/1983: 363; 1995). Even so, there is too little attention to the contested, political and chronic, stochastic uncertainty that pervades the entire process. Second, the notion of absorptive capacity is useful but insufficient. It is too mechanical and passive. Its usefulness is in suggesting that firms should anticipate future gaps in their expertise and capabilities. However, this notion lacks the 'cycle of learning' solution used by knowledge management theories. Third, knowledge management theories (e.g. Nonaka and Takeuchi 1995; Boisot 1998) suggest that the synthesized explicit knowledge inscribed in structured methodologies can be readily re-translated into local, embedded and explicit knowledge. They suggest a seamless elision from explicit dominated knowledge into tacit dominated communities of practice. In reality that politically redolent translation process is better grasped through PIEM (see Chapter 3) and similar politically nested approaches.

Appropriation occurs in an interactive and potentially emergent situation and should be understood as an active process. Domination, meanings and normative obligations are implicated and interwoven (Chapter 4) because a multiplicity of firms and parties are involved so there are issues of boundary spanning between different hybrid networks. Henderson and Clark (1990) suggest that the same innovation sold by the same supplier can be interfaced in two quite different ways to the user as:

(1) a customized configuration with attention to the architectural platform as well as the nesting of specific modules.
(2) A collection of standardized modules whose architecture is left to the different groups who purchase different modules (see Burcher 1991).

Boundary spanning and professional associations

DiMaggio and Powell (1991) contend that professional associations drive and deliver organizational innovation. They are therefore significant to the appropriation process. Swan, Newell and Robertson (1999) have examined the role of professional associations in seven nations and this is partly summarized in Table 9.1.

Medium and large-scale firms typically employ many specialists and outsource additional expertise. The most powerful professional associations include lawyers and accountants. They, like the medical profession, have strong linkages to the central state. These professions occupy a key role in both ingesting innovations and in their creative dissemination from consultancies, both national and global. Professionals of widely varying expertise occupy the technostructures of firms. These will include a rich collection concerned with the various strands of strategic co-ordination information technology systems. Among those involved in SCITS, one type of professional association has been extensively researched. This is the professional association of the American Production Inventory Control Society (APICS) with similar, though not equivalent, positions in the UK, Canada, Sweden, Germany, The Netherlands and France. These national associations affiliated to APICS possess formal channels down which the American centre can transfer modules of knowledge to the periphery. However, following on from our earlier analysis, it is likely that the peripheries will reinterpret the hub. Research suggests that there are firm effects from size and technology strategy as well as nation effects on the processes of networking.

Research in Europe indicates just how differently national users choose their packages. There are wide variations in the role of suppliers in authoring the body of knowledge. UK suppliers were the most active. They authored more than half of the articles in journals and two-thirds of the

TABLE 9.1 Inter-nation difference (derived from: Swan, J., Newell, S. and Robertson, M. 1999)

	CAPM systems adopted	Integration	Degree of customization	Perceived success	Using 'MRP2'	'MRP2' push by: PAs	'MRP2' push by: Suppliers
UK	Moderate	High	Moderate	Moderate	Moderate	Strong	Strong
France	High	High	Moderate	High	High	Weak	Moderate
The Netherlands	Moderate	High	High	High	Moderate	Strong	Moderate
Sweden	Very high	Low	Very high	High	Low	Weak	Weak

presentations at national conferences. Professional associations are shown to be perceived as being more comprehensive and impartial than the supplier and this indicates an important differentiation within the supplier–user interaction.

Within Europe there are clear differences in the approach to logistical systems between the Swedes, the Dutch, the French and the British, as shown in Table 9.1. The British are high in their concern for integration, but prefer standardized to customized modules. They perceive success to be moderate. The perception of high integration needs to be considered in the light of Burcher's (1991) demonstration that integrated systems barely existed and were not thought to be technically necessary. In the UK both the suppliers and the professional association are reportedly strong. The contrast is with the Swedish users who interact in a context typified by a weak push from the professional association and the suppliers. Swedish systems are very systemic and highly customized and with low integration. There is a strong perception of success, but this is coupled to a low perception that they are using MRP2. France and the Netherlands occupy intermediate positions between the Swedish and the British.

These professional associations played a recognizable role in boundary spanning between the suppliers and users for the production modality of this logistical process technology. However, this very insightful analysis of focused, one professional associations does not dis-aggregate the role of the accounting profession or of marketing. The system, however, implicates these as much as the production modality.

More than a decade ago there were clear indications of their growing role in British firms and their capacity to throw an over-layer on the accounting dimension rather than the production dimension. Since then the role of production relative to accounting and to marketing has become less significant and that probably reflects the tendency to mass customization coupled with the increasing role of surveillance and audits of performance.

Four episodes

DEF was constructed to direct attention to the key problems faced by user firms in appropriating innovations (Clark et al. 1992–93: 70). One aim is to combine Rogers' (1976) organizational adoption perspective with the innovation-design perspective that is oriented towards the interests of the user (Clark 1987; Clark and Staunton 1989: 200–205).

Figure 9.1 highlights the interwoven temporal sequence of four qualitatively different states in the articulations between the suppliers and the users through four major episodes. Each episode implicates different degrees and forms of uncertainty. Also, the episodes actively encounter the pre-existing power structure of the firm and the politicality of the heterogeneous players. It is presumed that different sets of human actors possess different conceptions of their interests and of their puzzle-solving regimes. This is shown as 'problem ownership' and the assumption is that the firm's division of expertise connects and assembles the distributed expertise of various relevant groupings and actors. Different groups and strata within the firm will be likely to lay claim to different domains of problem ownership (Clark 1972). Because of the division of expertise it is quite possible that different groups enter and leave the decision arena occupied by the suppliers. This element is well known to suppliers and many in the SCITS business insist upon the presence of end-users at an early state. However, in the public sector it is often the case that the early states are shaped by dominant professional groups even though nurses and auxiliaries operate the final application.

SCITS is a multi-layered innovation whose black-boxing is readily unpacked by the different groups in the user firm. This unpacking is also enabled by the suppliers.

The process of moving into and through the four states is therefore inherently conflictual. DEF suggests that the journey from the initial encounter between the user and the supplier through to the utilization state involves different combinations of heterogeneous actors. The firm's organizational repertoire and the degree to which there are recurrent action programmes for assembling an altering innovation represent the absorptive capacity. DEF acknowledges that organizations possess repertoires to cope with most operational units of analysis, such as the university year. There are other temporal units associated with innovation-design. In the saloon car sector those were five years and are now rather shorter. The repertoire for ingesting externally-based innovations is part of the absorptive capacity of the firm and will vary in its generative potential. The time frame will be project organization.

The notion of four episodes arose from the re-analysis of a five-year research consultancy project on organization design (Clark 1972) and was then used to analyse the innovation-design repertoire of British firms (Whipp and Clark 1986; Clark 1987; Clark and Starkey 1988). The four states represent the major categories used by the relevant groups and actors in the assemblage of their project. Once a project is underway the four episodes can be ongoing at the same time because of iterations and editing (Clark 2000).

Giddens defines episodes as 'sequences of change having a specifiable opening, trend of events and outcomes, which can be compared in some degree of abstraction from definite contexts' (1984: 374). This usage requires some modification to take account of the theory of structural activation.

The first episode is *agenda formation and design* of the organizational innovation. This episode relates to the process of initiation when certain boundary-spanning groups within the organization enter the solution-problem arena and engage in activities with external advisers and suppliers. This episode, which may be aborted, is an important initial shaping process, but its intended outcomes may be transformed in the next state. The suppliers will introduce their suggested solutions, the costs and their account of the implications. The definition of the problem and the solution will be mediated by the available organizational knowledge. It should be the case that the differences between the existing organizational processes and those intended for the future can be recognized, albeit in a fragmentary way that is complex for the in-house political system to process (Clark 1972). The differences can be a major area for scenario-building and for the PIEM process. Moreover, there are likely to be alternatives and each of these will possess differences that may require articulation.

The second episode is translation of the *design concepts and the initial selection* of the context-specific innovation. This involves the more detailed consideration of the proposed innovation and this might be unwittingly biased towards a technical fix as indicated in the vignette on the university. This should involve the reblending of the innovation. Decisions taken at this stage have implications for whether the evolution of SCITS in the specific context is an altering or entrenching innovation. Moreover, by this stage many groups are likely to be aware that changes in practice are underway.

The third episode is the *installing and commissioning* of modules and the overall architecture. Previous studies of organizational decision-making suggest that a huge variety of events may be unfolding in this episode and there are likely to be iterations back to edit and possibly alter the two previous episodes.

The final episode is *usage*. In the case of an analysis of a British car firm, there were dramatic differences between the organizational innovation that emerged and the intentions of those involved in the design-agenda episode and in the selection-translation episode (Whipp and Clark 1986). The emergent form revealed the complex interplay between structural conditioning and the different agentic intentions of the designers and the trade unions. Likewise, the example of the university mentioned earlier indicates that hybrids are the most typical.

Global Transfers and National Specificities 10

The success and failure rates for the global transfer of organizational innovations are unknown, yet a broad-brush estimate can be proposed. We know that a sizable minority of product launches fail. We can therefore estimate that at least 30 per cent of global transfers fail. Surveys of process innovations are rare, but the survey on Business Process Re-engineering suggests rates of failure as high as 60 per cent. However, for the moment we will run with the 30/70 split between failure/success even though the 70 per cent is probably too high. The broad brush suggests:

(1) Failure (more than 30 per cent)
(2) Varied forms of success and appropriation (less than 70 per cent)
   - Full hybrids (30 per cent)
   - Part hybrids: more than half the bundle of attributes adopted (20 per cent)
   - Outright reproduction (20 per cent)

Outright reproduction and imitation is much less frequent than theories of isomorphism suggest. Too much attention has been given to cosmetic similarities assessed through proxy variables and too little to the analysis of processes.

The transfer of organizational innovations between nations poses considerable problems. For a long time the transfer was treated as relatively unproblematic because modernist, pseudo-evolutionary theories such as those of Chandler (1977, 1990) dominated the discourse. These rightly emphasized the innovatory character of American organizing and its analytic repertoire (e.g. Whittington and Mayer 2000). Their conception of history and diagnostic relevance to the future is problematic (Clark 2000). For example, Dower's (1999) historical analysis of the relationship between Japan and the USA since 1853 shows the contextual dynamics of nations in the international political economy. However, the studies in the Chandlerian perspective often fail to examine the structural dynamics and position of American capitalism and management in the international political economy (e.g. Whittington and Mayer 2000). In addition, they invoke a too simplistic analytic repertoire for examining whether organizational innovations are similar. It is the use of organizational innovations as processes in action that matters and not cosmetic or stylistic similarities of structure.

se problems in the next section and then examine two
ie international trajectories of organizational innovations
l understood: global media sport and the retail com-
supermarkets. In each of these areas there are 'hidden'
tures and heterogeneous actor networks that enable the
he transfer process. However, hybrids abound. The final
on hybrids.

Problems of cross-cultural transfer

This section addresses the international and national levels of analysis
through three different frameworks, each of which is angled towards the
problems of cross-cultural understanding of process innovations. The aim
is to compare the approach derived from Hofstede (1980) with the relational configurational approaches.

First, an early speculative conceptual model by *Khedia and Bhagat*
(1988) makes strong use of the individualistic/systemic approach (Clark
2000) of Hofstede to develop propositions explaining the effectiveness of
transferring process innovations. They also propose some diagnostic rules.
Their framework takes four main sets of variables:

(1) The antecedent characteristics of the innovations.
(2) The antecedent differences between supplier and recipient organizations.
(3) The absorptive capacity of recipient organization.
(4) Cultural repertoires of the two nations are compared using the
dimensions proposed by Hofstede (e.g. power–distance, individualism
and masculinity).

This framework provides the basis for eight diagnostic propositions:

1. Process innovations that co-ordinate intra- and inter-firm activity are
more difficult to transfer than product technologies.
2. Process innovations may introduce and require new political configurations coupled to new truces and rewards. Therefore differences
in power–distance may be important, as in the case of the USA and
Japan.
3. If the sender and receiver are similar in antecedent conditions and
cultural values, then full transfer is more likely. Therefore differences
provide a strong challenge (e.g. UK/India on the transfer of coal-
mining technologies).
4. Individualistic cultures (e.g. USA) will be more successful in their
propensity to absorb and diffuse innovations than collectivist masculine cultures (e.g. Japan). This proposition fails to explain the
success of Japanese firms in establishing plants abroad.

5. Masculine cultures are more effective than feminine cultures in absorbing and diffusing.
6. Abstract cultures are more effective than associative cultures.
7. Differences in the negotiated orders between the two locations will adversely affect transfer.
8. Cosmopolitan organizations in societies with sophisticated technical and strategic orientations will be more successful than localistic organizations. This might partly explain the problems BMW experienced in their brief ownership of the Rover in Britain.

These propositions provide analysts with one framework for examining the interactions between different cultures and organizational innovations to locate the dynamics of absorptive capacities and the negotiated order. This is indicated by the next two examples.

Second, Guillen's (1994) synthesis is a bold attempt to calibrate the degree of adoption of major process innovations like Scientific Management and to explain both the degree and elements adopted with contextual factors. Guillen compares the degree of adoption as technique and ideology of three quite varied organizational innovations in the USA, Germany, Spain and the UK in a longitudinal investigation. The three are Scientific Management, Human Relations and structural analysis. Their adoption varies and Guillen draws on seven explanatory factors. This framework provides a neat contrast to the previous example.

Guillen concludes that in the period up to the 1960s Scientific Management was highly adopted in the USA and Germany as a technique but low as an ideology. In Spain and Britain there was very low adoption until after 1940. His explanatory factors are:

(1) *Corporate structure*. The structural changes in each nation in terms of five variables and their pace of evolution: size of firms (small/giant), ownership (private/public), Gross Domestic Product (low/high), complexity (low/high), bureaucratization (low/ high). The greater the size, public ownership, high GDP, high complexity and bureaucratization then the higher the adoption of Scientific Management.
(2) *International pressure*. The greater the international pressures the more the adoption. The USA scores low on pressure in part because of its low dependence on international trade. Germany was a late industrializer, became politically isolated and viewed the USA as a threat.
(3) *Industrial relations system*. The USA had adversarial conflict by very powerful management against powerful unions. Great Britain had delegated shop-floor control, high labour unrest, and craft-based unions. In Germany, labour unrest was inhibited by reformist unions, by corporatism and bureaucracy.
(4) *Mentalities of business elite*. American elites were typified by progressivism, a craze for efficiency, technocratic Modernism, cultures

of mass production and then mass customization. German elites were modernistic. British elites were pro-market mechanisms rather than the visible hand (see Chandler 1990).

(5) *Professions*. American professions were efficiency experts with a powerful role for the engineering professions. Germany also has numerous highly-trained professional engineers occupying powerful positions in firms. Britain, like Spain, had few qualified engineers and no strong professional associations. Many engineers were sceptical about Scientific Management.

(6) *State role*. The American state was weak and initiatives were located in the private sector. The German state was strong (e.g. National Board of Efficiency). The British state was weak and state research was anti-Scientific Management and praising Human Relations.

(7) *Workers response*. In the USA there was strong shop-floor resistance while German unions promoted Scientific Management. Spanish and British unions were also opposed.

These seven factors differ from those favoured by Khedia and Bhagat. In the Hofstede framework the British and Americans are closer to each other than to either the Germans or the Spanish. However, in each of the three organizational innovations studied there are sharp Anglo-American differences in the form and degree of adoption.

Third, conventional narratives fail to situate biographies and social histories. This example from Australia evokes the locality and placeness of organizational innovation. Peppertree is a hotel, luxury restaurant and winery in Australia's most visited wine-tourism district. *O'Neill and Whatmore* (2000) employ actor network theory to construct three pathways through the Peppertree network. These networks are inherently unstable yet act as performative agents in relations and organize the spatial imaginaries. The three pathways reveal the interweaving of biographies and social relationships to show innovation as a complex, social, endogenous and collaborative network:

(1) The first pathway is about the four married couples centrally involved. Their intimacy, hard times and the lucky breaks. Their partnership was a complex division of knowledges and power relations that was intimately gendered through the social connections and personal relations of the heterosexual couples. The cadences and dynamics of marital relations intertwine with the legal and fiscal arrangements. Their social networks connecting them to social elites in Sydney are an important currency in the marketing and customer knowledge vital to performance (O'Neill and Whatmore 2000: 128).

(2) The second pathway focuses on the convent building that was transformed into Peppertree. This building was central to the local civic sense of its community. Locals were agitated about the proposal to demolish the building for new homes. The building was bought by the partnership, taken apart and then transferred to its new site at

Peppertree. The convent was a heterogeneous assemblage whose future resided in the social inertia of its original design as an aesthetic spatial imagery. The convent was re-located in a new spatial imagery 'to create a *cuisine de terroir* in which climate and soils and the cadences of plant and animal life are drawn into the Peppertree network' (ibid. 2000: 132).

(3)   The third pathway is the gastronomic landscape of food, *terroir* and the local ecology. The transferred convent becomes the restaurant – Roberts at Peppertree – and it provides the impression as socially constructed by its elite consumers of a 'large airy Australian settler cottage with massive Australian hardwood rook structures and floors. It has an "outdoor feel"' (ibid. 2000: 132). The design intensifies the connections with the landscape and the local ecology of food production by anchoring food production. The partners had access to the considerable networks of local fresh suppliers in Hunter Valley. Peppertree appealed to the sense of identity and sacredness for the post-colonial Australians from the Sydney and related elites.

This piece exemplifies the context dependency of organizational innovations and technology as process. Its attention to the biographies and social history implicated in organizational innovations has considerable relevance.

Global-media sportization

The example of sportization traces the links between innovation, sport, globalization and state formation in three steps:

(1)   the Elias–Dunning thesis of sports and society.
(2)   Maguire's model of the evolution of sports from the sixteenth century onwards.
(3)   The contemporary growth of global media sport.

First, Elias and Dunning theorize sports as highly regulated contests and cultural forms developing from the sixteenth century period of conflict among western European nations. Elites in those nations sought to extend the emotional pleasure of victory in real wars to a mock battle. This commenced at the founding moments of the modern period. Sports are hypothesized to arise from a changing balance of power over several centuries, especially in the English context. In eighteenth-century England the three-way struggle of elites between the aristocracy, the gentry and the rising bourgeoisie contained tension-balances that socially conditioned the psychological needs for spontaneous, basic, unreflective, pleasurable excitement. Elias contends that since then the control of excitement in public settings (e.g. public hangings) and hence of personality structure has involved a progressive narrowing of acceptable levels of excitement and violence. In the eighteenth century various sporting and leisure activities surfaced. It is postulated that these activities counteracted and interacted with 'stress tensions' in the tension-balance between the aristocracy, gentry and bourgeoisie. These tensions provided different opportunities for the legitimate, controlled decontrolling of emotions. So, within the English public schools activities that later became known as soccer and rugby union emerged within the different settings of the aristocratic Eton School (soccer) and lower elite strata of gentry-dominated Rugby School (rugby union). These sports provided for different moods and pleasurable excitement.

From this process and configurational perspective I would draw two important implications:

(1)   Contests played out in sporting activities have a 'mimetic relationship' to the overt inter-nation struggles and are as equally real as any other area of social life.
(2)   Sport, as a social container is the carrier of the deep cultural and structural repertoire because sport is encoded and suffused with sedimented selection criteria from the founding configuration. Therefore, the different ways in which sports are co-ordinated is significant.

Clark (1987, 2000) specifically contends that American Football embodies and articulates the deep culture and structure of the American male service class in the post-bellum period of the American Civil War. That suggests that sports are potentially a powerful mechanism for transferring strata-based scripts such as the notions of teamwork and of how to organize (Clark 1987: Ch. 6). Once established, each sporting activity acquires a degree of autonomy.

Second, as shown in Figure 10.1, Maguire (1999) presents a preliminary model of the varieties of sportization in five historical phases:

(1)   In first the period from 1500 to 1750 there was an Anglo-European core. In England, after the 'calming down' of the violence of the Civil War (mid-seventeenth century), Parliament became the symbolic battleground for the aristocracy and the gentry. Military skills on the battlefield were replaced by verbal skills in Parliament.
(2)   In the 1750–1870s period the tension-ridden context and social habitus of the English ruling class sustained a variety of processes that became modern sports. There were two main waves:

   •   In the eighteenth century the sports were fox hunting, horse racing and boxing
   •   In the nineteenth century the sports were soccer, tennis, rugby and athletics.

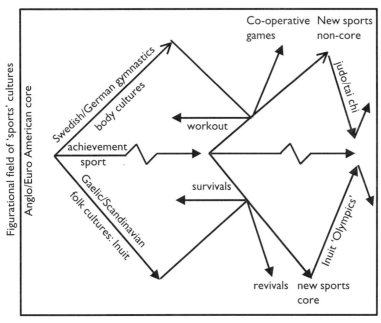

FIGURE 10.1   Global sportization (from: J. Maguire 1999)

The landed aristocracy and gentry dominated the 'free' peasantry and were able to patronize and appropriate folk pastimes. Their approach to these male-dominated pastimes involved a distinct body culture and set of values.

(3)   In the nineteenth-century American context the male body culture was significantly oriented towards achievement and winning. This was especially so in the period from the 1870s to the 1920s (Maguire 1999: 82–88).

(4)   During the period from the 1920s to the 1960s there was an American influence with the diffusion of baseball, basketball, ice hockey and volleyball within the American sphere of influence (cf. Europe). American Football did not diffuse to the same degree as association football. American male versions of sport practice and of the achievement sport ethos became ascendant.

(5)   Between the 1960s and 1990s there was the rise of televised spectacles: theatres of consumption. American marketing strategies became highly visible. Moreover, American approaches to management, marketing, organization and choreographing were very evident, although not yet a template for the rest of the world. During the Cold War sports were aligned with national symbols. Soviet bloc nations and their sphere of influence challenged western hegemony, especially in the four-yearly Olympics. Sports were special space–time arenas of analytic expertise. In the west, sport was increasingly commodified

and became part of the cultural industries. Commodification for television was often enabled by the advertising industry searching for consumers to enrol for their brands. The advertisers were certainly interested in marketing sports goods (e.g. Nike) and were also interested in the visibility that could be given to particular signs of consumption (e.g. cigarette-making). Parallel to the interest in the major established sports there was a growing interest in niche sports and markets, such as surfing, hang-gliding and a range of 'peak experience' extreme sports. At the start of the third millennium – and in the late fifth period – there is not only a contest between nations but also a figuration involving contest between social groups. Within that figuration particular attention should be given to the global media–sport complex and to the commodity knowledge-chains in sport.

Third, Maguire (1999) shows that the growth in importance of the global media–sport complex has been dramatic because viewers of television are consuming ever more global media-sport. Maguire observes that it is very big business in terms of television, advertising, merchandising and the sale of exclusive rights for events, spectacles, tournaments and even leagues (1999: 144f.). The global media–sport complex is a subset of the overall process of globalization and draws therefore on the forms of new knowledge contained in its business service organizations. The three main parties in the complex – sports organizations, media marketing groups and the transnational organizations – are very powerful.

Global media–sport is tightly interwoven with global media communications, as in the case of the Australian entrepreneurs, Murdoch and Packer. Therefore the contemporary consumer experience of sport is intertwined with the interests and actions of the global media. In the USA by the mid-1960s the television consumer had replaced the live consumers as the focus of capitalism. It is the consumers' lives that are being commodified in front of the television screens. The sporting events are relatively cheap to show given the large audiences of viewing consumers that are made available to the advertisers.

The global media–sport complex embraces a diverse array of flows and interdependencies that are part of overall globalization. The flows include personnel, capital, technologies and a complex collection of corporate and national symbols. These flows are a disjunctive system (Appadurai 1986b) typical of re-organized capitalism (cf. Lash and Urry 1987). The interdependencies involve both the creation of a specialized subset of the transnational capitalist class and a closely associated yet relatively autonomous collection of transnational practices. These transnational practices possess a relative autonomy and contain the sedimented selection criteria that structure the future actions and plans of the capitalist strata. The global media–sport complex is permeated with implicitly American codes, especially regarding specialized knowledge and practices about orchestrating the consumers' consumption of televised sport and related products. The character of ownership is more criss-cross in the USA than in Europe.

In Europe between 1989 and 1995 annual media–sport coverage more than doubled. From 1995 into the new millennium the coverage doubled again according to the European TV Sports Databook (Kagan World Media 2000). However, we still know remarkably little about how the consumer consumes televised sport and how the content is experienced and utilized. Even so, the figuration of knowledge around global media–sport promises those investing corporations very favourable returns to future sales revenue. Sport has been commodified so that it is the size and composition of the media audience available to advertisers and sponsors that determines the value of particular sports.

Increasingly, clusters of parties are socially constructing sports to be revenue-providing entities. The investors hope to emulate the trajectory created by Eccleston with Formula-1 motor racing and to establish their chosen commodity as a global media–sport that they own. The media–sports–knowledge complex is about enhancing the excitement and spectacle of the revenue-generating elements in sports. Media–sports personnel are a new occupation of symbolic analysts comparable to the imagineers at Disney. They play a crucial role in producing marketable commodities that possess semiotic materiality, that is, they intermingle visual, audio, feeling and other codes in addition to the written and engage the soul of the consumer. Because the souls of the global consumer cut across so many identities and boundaries of gender, ethnicity, religion and nationalism, there is no single monolithic ideology. Therefore, in global Formula-1 motor sport the image of masculinity is an intriguing one! The Formula-1 complex of vested interests holds a monopolistic position relative to other European competitors.

Global retail commodity chains and performative knowledge

The time–space stretching and connections that link producers and consumers have been explored in a number of ways. Previously, the global commodity chain was a reductionist and linear framework that omitted elements central to examining the role of knowledge. The chain tended to be a-contextual and over abstracted in a way that imprisons analysis and reduces local variety. There was insufficient attention to examining the causal mechanisms through which chains are constituted and undermined. This section examines the globally-driven chains of British supermarkets sourcing horticultural products in the Third World.

The notion of the global commodity chain has been reformulated by Gereffi (1994) as a network or chain that links all the agents from final distributor back to the early stages of production and is governed unequally by the more powerful global agents in the network. The novel feature is that at the global level within each chain a very small number of firms are making strategic decisions and constructing the economic networks that connect the spread of economic activities across the world.

These pivotal firms are strategic brokers in making the chains. The global firms are organized in one of two types of governance structure in relation to the chains: the producer-driven chains or the end-buyer-driven chains. The producer-driven commodity chains are typical of capital and technology-intensive industries where the highest barriers to entry are located in the core technologies (e.g. automobile sectors). The end-buyer-driven chains are those in which retailers and brand-name companies play the governing role (e.g. The Gap, Nike). Frequently, these firms do not own facilities (e.g. Ikea). They concentrate upon innovation-design and distribution in the market. These firms rely upon complex multi-tiered networks that they have constructed. Profits for the end-buyer derive from the unique combination of high-value design and development drawing upon varied knowledge bases. They are engaged in brand-name development (e.g. Liz Claiborne).

The pivotal organization that leads the chain is continually engaged in four network building activities:

1   Attracting and shaping new members.
2   Sustaining the network by managing conflicts and enabling learning. Building the culture of the network. Establishing its governance systems and shaping its structure.
3   Positioning their network in the global market.
4   Creating durable and information-rich relationships.

These chains are hybrids containing elements from both arms-length contracting and obligational contracting. This can be illustrated from the sourcing of horticulture in the Third World by leading British supermarkets. The supermarket buyers have brand-names with power in the British market but not abroad. They have developed logistical systems that are tightly connected and have immense consequences for their suppliers. How are these chains governed? The British chains are dominated by a few large supermarkets that have chosen to compete through fresh fruit and vegetables by presenting these on a year-round basis with strong aesthetic appeal. There are wide product ranges with continuous innovation and including the provision of recipes. Also, there is an articulation of this presentation with the media scapes articulated through the many television programmes on cooking and household design. This may explain why fresh produce is one of the destination categories for which shoppers switch stores. Their sales have grown considerably. The chains have strategically organized their knowledge to address six dimensions:

•   Quality and consistency of visual appeal, shape, texture and flavour
•   Reliability of supply with availability round the year
•   Costs are important and supermarkets seek a reduction in transaction costs through scale
•   Variety, added value and innovation

- Food safety and due diligence
- Ethical trade (e.g. not exploiting Third World employees).

Specialty vegetables in British supermarkets show considerable variety. This informs how there is re-structuring of the vegetable chain into very sophisticated supply chains.

There are two major routes to transforming earlier chains. First, many intermediaries have been eliminated. Second, the domestic Third World industry has been transformed. For example, British supermarkets have constructed a pivotal role in which their strategy is to define the outputs of the chain and who is in the chain, as well as defining the distribution of activities between different agents. They also monitor the performance of the chain. The redistribution of activities illustrates the role of the supermarkets. The major supermarkets all concentrate upon their core role of retailing activities of branding, product innovation, product design and marketing. They search for ways of reducing the costs and the risks of procurement, processing and quality to themselves by locating these problems with the other agents. Major activities have been transferred from the UK to the Third World, including the washing and trimming of vegetables as well as bar coding and labelling. This means that the suppliers have to acquire considerable organizational capabilities. Suppliers are responsible for the overall management of the post-harvest activities, including achieving consistency in the cold storage chain. This is crucial because a one hour delay in removal from the growing field can reduce shelf life by eight hours so products must be placed in a cool place after harvesting and then transported in humidity-controlled conditions. These require considerable investment and capabilities in handling complex machinery.

This example of the commodity chain highlights the relative power position and strategic choices of the end-buyer while showing that Third World suppliers occupy limited zones of manoeuvre.

There are features of how knowledge in the chain is organized that remain obscure. An alternative theorizing to the global commodity chain could reduce the linear assumptions of the global commodity chain and *accentuate a network metaphor* (Hughes 2000). This should include:

- Non-linear circuits without starting and ending points
- Webs of interdependence (Powell, Koput and Smith-Doerr 1996)
- Contextual understanding
- How meanings get moved around
- Identification of the various cultural knowledges.

Earlier research had focused narrowly upon the production rather than the consumption spheres and tended to apply a narrow definition of knowledge, excluding aesthetic knowledge. More recent research has exploded the consumption sphere by exploring the geographical settings and changing cultural biographies of commodities (see Chapter 6). There are considerable implications of this in revising the global commodity chain to separate production and consumption while allocating an explanatory role to mediations. Too many studies have omitted consumption and the issue of mediating roles. The mediating role of the retailer uses 'maps' and SCITS to achieve the assemblage of diverse knowledges and to create knowledge-using systems that may strengthen and lengthen networks. Inflecting 'commodity cultural knowledge' into the network metaphor and exploring the mediations between production and consumption can give richer analysis. There are strong possibilities for exploring the constitution of flows and the circuits of knowledge.

The notion of circuits of performative knowledge amplifies Gereffi's global innovation networks. In the global networks for delivering cut flowers to the customer, the mediating role of the supermarket has displaced rival retail outlets and is re-structuring the whole global innovation network (Hughes 2000). The supermarkets are bypassing the role of previously significant agents, including importers, Dutch auctions and various warehousing services. Supermarkets have reduced the length of time in the chain in order to minimize the perishability of the cut flowers and extend their selling and viewing durations. They continue to redefine the design and aesthetics of the commodity. The supermarket has come to occupy a mediating role in the capitalizing of the social, technical and aesthetic knowledges used by producers and distributors. Retailers are developing performative knowledge in innovation-design by translating consumer research into the design and aesthetic specification that is used to manipulate the floricultural (genetic) knowledge applied to the production process. Their managerial strategies fuse the knowledge of the consumer for the retailer and for the producer with a circuit of knowledge based on focus groups, panels and questionnaires. The supermarkets initiate innovations in bouquet design through their dialogue with top-level specialist designers such as Jane Packer. This is a circuit in design-based ideas that connects expert knowledge about consumption media scapes to the expert knowledges of floriculture. Retailers enrol floristry experts and other agents in the circuits of design-based knowledge. There are also circuits that take up the technical possibilities on florigenetics and create in advance of production the knowledge necessary for production through design-and-development. Retailers are drawing together top florists and experts in florigenetics in networks that contain complex, multi-stranded relationships between a diverse range of agents. There are circuits of social, technical and aesthetic knowledges within the global innovation networks. This is a knowledge-intensive network between producers and consumers via the mediating role of retailers that incorporates:

- Technical knowledge (florigenetics, field experimentation, refrigeration and cooling systems)
- Aesthetic knowledge (cultural meaning of flowers, colour, display)
- Social knowledge (gift occasions, identity, social practices of display).

The three types of knowledge are distributed across a variety of centres and get connected and pushed around by the retailers. The retailers' action in circulating the knowledges is based on their managerial knowledge about competition and consumption. The capitalizing of the retailers' knowledge depends on the organizational capacities to reformulate the relationships between the nodes. The role of knowledge circuits affects the capacity, duration and strength of the networks. Therefore the strategic management of the networks involves the division and distribution of knowledges. With the influence of the retailers there is an extension to the knowledge centres of the florigeneticists: breeding flowers, the cultivation process, the development of pesticides, cross-breeding to produce different colours, creating new varieties and extending the vase life. The retailers are in a strong position to capitalize on the circulation of the knowledges between different centres on the network and to strengthen their position accordingly. They are employing power-laden commercial capital. Clark and Staunton (1989) contend that there is an investment in categories. The retailers' puzzle-solving regime is a core competence that is expensive and sets the strategic directions of the firm.

Hybrids, economies of scope and customization

The current, third research programme (as Chapter 2) seeks to problematize the earlier accounts of transferring organizational innovations between sites in one nation and then from there into the global context. In the degenerating research programmes (Lakatos 1978) there is an impression of successful transfer. The impression has been sustained by the elegant and informative style of key narrators like Chandler and Rogers. They have provided the props and scripts that others have eagerly performed. Rogers' strong analytic framing of the diffusion of innovations has been immensely influential and that framing infused the isomorphism hypothesis. The discourse from these earlier research programmes created a strong sense of organizational innovations being secondary to technology artifacts and as being readily transferable as a whole by competent managers applying expert modernist knowledge. Even the new Resource Based Theory is preoccupied with how firms can protect themselves from imitation (Grant 1998). Those earlier research programmes are losing their protective belts.

In the current research programme there is a reformulation of the background assumptions, the metaphors, the analogies and the conceptual repertoire. This reformulation is neither focused in particular centres nor confined to the established discourse of North American versus European approaches. The reformulation is rather dispersed and yet is gaining ground in both the established and new journals. Moreover, the reformulation is even better understood in corporations than in the universities.

The new political economy research programme rightly adopts a multi-level approach to organizational innovation because the global level has not been sufficiently inscribed in earlier research programmes. The multiplicity of levels and the configurations (Chapter 4) that emerge illustrate both the centrality of the state mechanisms and also the *ineffectiveness of architectural policy-making*. It is unlikely that the state can 'design' the conventions of co-ordination that are in the national repertoire (Storper and Salais 1997). It is therefore unlikely that the polity and bureaucracy of the European Union can simply legislate by articulating a European Innovation Monitoring System (EIMS).

The four ensembles of Storper and Salais examined in Chapter 6 usefully interface with the hypothesis of mass customization by highlighting the persistence of forms reflecting mass production, yet also highlighting the variety of alternatives. The concept of heterogeneous engineering indicates just how complex are the problems of co-ordination in the world of mass customization. The implication is that firms will have to be highly differentiated into multiple sub-units learning specific processes while that differentiation requires high integration. SCITS are playing a major role as Castells (1996) suggests, yet the elective affinity between the potentials of their core dimensions may prove as much beyond the cultural repertoires of some nations as was Scientific Management (Guillen 1994). In the new political economy some nations might construct complementary conventions of co-ordination through the action of their firms. This was the situation between 1965 and 1985 for Japan and the USA (Clark 2000: Ch. 9) and J–USA is a useful instance of multiple-nation configurations that prove the necessity to reformulate Porter's diamond framework on innovation and upgrading.

The limitations of transfer because of contextual features and dynamics are increasingly recognized. For example, policy approaches in geography to regional innovation increasingly emphasize the limits to learning (e.g. Hudson 1999). So far, these contextual processes have been under-analysed in organizational approaches.

There is now both widespread evidence of hybridization and also an increasing analytic commentary on the generative mechanisms involved. The difficulties of becoming global are obviously affecting European firms. Some have simply found that their profitable activities have been confined to their domestic markets (e.g. Marks & Spencer; cf. Ikea). The German core of the European automobile industry is revealing of the different outcomes. In Europe Volkswagen are the market leaders and have successfully re-branded the Skoda marquee. BMW failed to transform the rump of the British automobile industry. Some journeys are a brand too far. Arguably their strategies of organizational innovation were rooted in romanticism about British forms of organizational innovation. The Mercedes-Chrysler project defies a clear conclusion at this moment. This mix of self-imposed experiences illustrates the problems of stretching space–time and of ignoring heterogeneous engineering. Equally, American and British firms entering Germany have found the regulative context over space–time confined and initially confounded their core heuristics about global success (e.g. Wal-Mart and Virgin). Yet the case of McDonald's confirms that some global firms possess elites that can de-code new national contexts and can develop their own infrastructures. McDonald's reportedly spent ten years establishing their British network of franchises. McDonald's have both customized their services to national tastes and shown awareness that it is their process organizing which is the heartland of profitable innovation. The investment in the British market leader in mid-market sandwiches (Pret à Manger) at its moment of entry in New York and the USA suggests the repertoire is being extended in scope and richness. Somewhat in contrast, Service Corporation International, America's leading bereavement supplier over-estimated the value of indigenous firms when moving into the global market. They are divesting from some nations while retaining their overseas platforms in Australia and Britain. Moreover, their recent fiscal crisis has created a channel for those with overseas experience to move back to the USA into more central positions in order to confront their engagement with mass customization.

Conventions of co-ordination are heavily embedded and articulate with the national habitus of the domestic corporate elites (Storper and Salais 1997; Clark 1997). Even allowing for each nation to possess a typical variety rather than a single pattern of co-ordination conventions, it is very likely that for most nations the elective affinity of hybridizing will not coincide with performance dimensions for success (Best 1999; Castells 2001). The British case shows many sectors in which firms from overseas have parented and co-ordinated the domestic workforce more effectively than the British service class. Examples include fast food, automobiles (e.g. Peugeot, Ford, GM, Toyota, Nissan, Honda), leisure, bereavement, high-tech services and others.

The USA/Europe interface is especially relevant. Few expected that American exceptionalism included capacities for liminality in existing sectors while providing key spaces for new sectors. Storper and Salais (see Chapter 6) suggest that the American conventions of co-ordination through the market are so dominant in the discourse and the social capital that the possibilities for mass customization are awkward. However, specialists in global production methods like Schonberger (1981) and Pine (1993) recognized that the USA contains capacities for both entrepreneurialism and for standardized repetitive firms (Castells 1996). Their contributions to understanding co-ordination have largely been sidelined until the current research programme. Abo (1994) and colleagues have carefully examined the Japanese production system in the USA and concluded that the emergent outcome is the hybrid factory. Their reasoning provides a careful refutation of the notion that Japanese systems are the new universal solution (cf. Kenney and Florida 1993). The hypothesis of hybridization fits well with the under-used concept of elective affinities discussed earlier. *Hybridization should be the hypothesis to be explored.*

References

Abbott, A. (1988) *The System of Professions. An Essay on the Division of Expert Labour*. Chicago: University of Chicago Press.

Abbott, A. (1992) 'From causes to events: notes on narrative positivism', *Sociological Methods and Research* 20: 428–455.

Abernathy, W.J. (1978) *The Productivity Dilemma. Roadblock to Innovation in the Automobile Industry*. Baltimore, MD: Johns Hopkins University Press.

Abernathy, W.J. and Clark, K.B. (1985) 'Innovation: mapping the winds of creative destruction', *Research Policy* 14: 3–22.

Abernathy, W.J. and Hayes, R.H. (1980) 'Managing our way to economic decline', *Harvard Business Review* September/October: 69–77.

Abernathy, W.J., Clark, K.B. and Kantrow, A.M. (1981) 'The new industrial competition', *Harvard Business Review* September/October: 68–81.

Abernathy, W.J., Clark, K.B. and Kantrow, A.M. (1983) *Industrial Renaissance: Producing a Positive Future for America*. Boston, MA: MIT Press.

Abo, T. (ed.) (1994) *Hybrid Factory. The Japanese System in the United States*. Oxford: Oxford University Press.

Abrahamson, E. (1991) 'Managerial fads and fashion: the diffusion and rejection of innovations', *Academy of Management Review* 16: 588–612.

Abrahamson, E. (1996) 'Technical and aesthetic fashion', in B. Czarniawska and G. Seven (eds), *Translating Organizational Change*. Berlin: de Gruyter, pp. 117–137.

Abrahamson, E. (1966) 'Management fashion'. *Academy of Management Review*, 21: 254–285.

Abrahamson, E. and Fairchild, G. (1999) 'Management fashion: lifecycles, triggers, and collective learning processes', *Administrative Science Quarterly* 44: 708–740.

Abrahamson, E. and Fairchild, G. (2000) 'Who launches management fashions? Gurus, journalists, technicians or scholars?', in C.B. Schoonhaven and E. Romanelli (eds), *The Entrepreneurship Dynamic in Industry Evolution*. Stanford, CA: Stanford University Press.

Abrahamson, E. and Fombrum, C.J. (1992) 'Forging the iron cage: interorganisational networks and the production of macro-culture', *Journal of Management Studies* 29: 175–194.

Ackroyd, S. (1995) 'On the theory of organisational constitution and societal structuration', in H. Bouchiki, M. Kilduff and R. Whittington (eds), *Action, Structure and Organizations*. Warwick: Warwick Business School Research Bureau, pp. 1–30.

Ackroyd, S. and Fleetwood, S. (eds) (2000) *Realist Perspectives on Management and Organisations*. London: Routledge.

Akrich, M. (1992) 'Beyond the construction of technology: the shaping of people and things in the innovation process', in M. Dierkes and U. Hoffman (eds), *New Technology at the Outset. Social Forces in the Shaping of Technological Innovations*. Frankfurt: Campus Verlag, pp. 173–190.

Albrow, M. (1996) *The Global Age*. Cambridge: Polity Press.

Aldrich, H. (1979) *Organizations and Environments*. Englewood Cliffs, NJ: Prentice-Hall.

Aldrich, H. (1999) *Organizations Evolving*. London: Sage.

Alexander, J.C. (1995) *Fin-de-Siècle Social Theory: Relativism, Reduction, and the Problem of Reason*. London: Verso.

Ambler, T. and Styles, C. (2000) *The Silk Road to International Marketing. Profit and Passion in Global Business*. Edinburgh: Pearson Education.

Anderson, P. (1998) *Origins of Postmodernity*. London: Verso.

Appadurai, A. (1986a) *Modernity at Large. Cultural Dimensions of Globalization*. Minneapolis, MN: University of Minnesota Press.

Appadurai, A. (1986b) *The Social Life of Things. Commodities in Cultural Perspective*. Cambridge: Cambridge University Press.

Archer, M. (1995) *Realist Social Theory: The Morphogenetic Approach*. Cambridge: Cambridge University Press.

Archer, M. (1996) 'Social integration and system integration: developing the distinction', *Sociology* 30 (4): 679–699.

Arrighi, G. (1994) *The Long Twentieth Century. Money, Power and the Origins of our Times*. London: Verso.

Arrighi, G. (2000) 'Globalization and historical macrosociology', in J. Abu-Lughud (ed.), *Sociology for the Twenty-First Century. Continuities and Cutting Edges*. Chicago, IL: University of Chicago Press.

Arthur, W.B. (1989) 'Competing technologies, increasing returns, and lock-in by historical events', *Economic Journal*, 99 (193): 116–131.

Ashby, W.R. (1956) *Introduction to Cybernetics*. London: Chapman and Hall.

Attewell, P. (1992) 'Technology diffusion and organizational learning: the case of business computing', *Organization Science* 3: 1–19.

Bale, J. (1994) *Landscapes of Modern Sport*. Leicester: Leicester University Press.

Barley, S.R. (1986) 'Technology as an occasion for structuring: evidence from observation of CT scanners and the social order of radiology departments', *Administrative Science Quarterly* 31: 78–108.

Barley, S.R. (1990) 'The alignment of technology and structure through roles and networks', *Administrative Science Quarterly* 35: 61–103.

Barley, S.R. and Kunda, G. (1992) 'Design and devotion: surges of rational and normative ideologies of control in managerial discourse', *Administrative Science Quarterly* 37 (3): 363–369.

Bartlett, C. and Ghoshall, S. (1989) *Managing Across Borders. The Transnational Solution*. Cambridge, MA: Harvard Business School Press.

Baudrillard, J. (1975) *The Mirror of Production*. St Louis, MO: Telos Press.

Baudrillard, J. (1983) *Simulacra and Simulations*. New York: Semiotext.

Baudrillard, J. (1988) *America*. London: Verso.

Bauman, Z. (1992) *Intimations of Postmodernity*. London: Routledge.

Bauman, Z. (2000) *Liquid Modernity*. Cambridge: Polity Press.

Bell, R.M. (1978) *Changing Technology and Manpower Requirements in the Engineering Industry*. London: Sussex University Press.

Benjamin, W. (1999) *The Arcades Project*. Cambridge, MA: Belknap.

Beniger, J.R. (1986) *The Control Revolution. Technological and Economic Origins of the Information Society*. Cambridge, MA: Harvard University Press.

Bennis, W., Benne, K. and Chin, R. (1961) *The Planning of Change*. Holt, Rinehart & Winston.

Berger, S. and Dore, R. (1996) *National Diversity and Global Capitalism*. Ithaca, NY: Cornell University Press.

Best, M. (1990) *The New Competition. Institutions of Industrial Restructuring*. Cambridge: Polity Press.

Best, M.H. (1999) *The Work of Nations. Preparing Ourselves for the 21st Century*. New York: Vintage.

Bhaskar, R. (1975) *A Realist Theory of Science*. Leeds: Basic Books.

Bijker, W.E. (1995) *Of Bicycles, Bakelites, and Bulbs: Toward a Theory of Sociotechnical Change*. Cambridge, MA: MIT Press.

Bijker, W.E. and Law, J. (1992) *Shaping Technology/Building Society. Studies in Sociotechnical Change*. Cambridge, MA: MIT Press.

Bijker, W.E., Hughes, T.P. and Pinch, T. (eds) (1987) *The Social Construction of Technology*. Cambridge, MA: MIT Press.

Blackler, F. (1995) 'Knowledge, knowledge work and organizations: an overview and interpretation', *Organization Studies* 16 (6): 1021–1046.

Boisot, M.H. (1987) *Information and Organizations. The Manager as Anthropologist*. London: Fontana.

Boisot, M.H. (1995) *Information Space. A Framework for Learning in Organizations, Institutions and Culture*. London: Routledge.

Boisot, M.H. (1998) *Knowledge Assets: Securing Competitive Advantage in the Information Economy*. Oxford: Oxford University Press.

Boltanski, L. (1987) *The Making of a Class: Cadres in French Society*. Cambridge: Cambridge University Press.

Boltanski, L. and Thevenot, L. (1988) *Les économies de grandeur*. Paris: Presses Universitaires de France.

Boulding, K.E. (1956) *The Image*. Ann Arbor, MI: University of Michigan.

Bourdieu, P. (1977) *Outline of a Theory of Practice*. Cambridge: Cambridge University Press.

Bourdieu, P. (1984) *Distinction: A Social Critique of the Judgement of Taste*. Cambridge, MA: Harvard University Press.

Bourdieu, P. and Wacquant, L.J.D. (1992) *An Introduction to Reflexive Sociology*. Cambridge: Polity Press.

Braudel, F. (1982) *The Wheels of Commerce. Civilisation and Capitalism in the 15th–18th Century*. London: William Collins.

Braudel, F. (1985) *The Wheels of Commerce. Volume II. Civilization and Capitalism, 15th–18th Century*. London: Fontana Press.

Brown, J.S. and Duguid, P. (1991) 'Organizational learning and communities-of-practice: toward a unified view of working, learning, and innovation', *Organization Science* 2: 40–57.

Buckley, W. (1967) *Sociology and Modern Systems Theory*. Englewood Cliffs, NJ: Prentice-Hall.

Burcher, P. (1991) 'The use of capacity requirements planning in manufacturing planning and control systems', unpublished doctoral thesis, Aston University.

Burns, T. and Stalker, G.M. (1961) *The Management of Innovation*. London: Tavistock.

Burrell, G. (1992) 'Back to the future: time and organization', in M. Reed and M. Hughes (eds), *Rethinking Organization: New Directions in Organization Theory and Analysis*. London: Sage.

Burt, R.S. (1992) *Toward a Structural Theory of Action*. New York: Academic Press.

Burt, R.S. (1992) *Structural Holes: The Social Structure of Competition*. Boston, MA: Harvard University Press.

Butsch, R. (2000) *The Making of American Audiences. From Stage to Television, 1750–1990*. Cambridge: Cambridge University Press.

Callon, M. (1986) 'Some elements of a sociology of translation: domestication of the scallops and the fisherman of St Brieue Bay', in J. Law (ed.), *Power, Action and Belief: A New Sociology of Knowledge? Sociological Review Monograph 32*. London: Routledge.

Callon, M. (1997) 'Actor–network theory – the market test', Working Paper. Paris: Ecole des Mines de Paris.

Callon, M., Law, J. and Rip A. (eds) (1986) *Mapping out the Dynamics of Science and Technology: Sociology of Science in the Real World*. London: Macmillan.

Campbell, C. (1996) 'The meaning of subjects and the meaning of actions: a cultural note on the sociology of consumption and theories of clothing', *Journal of Material Culture* 1: 93–106.

Campbell, D.T. (1965) 'Variation and selective retention in socio-culture evolution', in G.I. Barringer and R. Mack (eds), *Social Change in Developing Areas*. Cambridge, MA: Schenkman, pp. 19–49.

Carlstein, T., Parkes, D. and Thrift, N. (1978a) *Making Sense of Time*. London: Edward Arnold.

Carlstein, T., Parkes, D. and Thrift, N. (1978b) *Timing Space and Space Time*. London: Edward Arnold.

Carrier, J.G. and Miller, D. (1998) *Virtualism: A New Political Economy*. Oxford and New York: Berg.

Castells, M. (1989) *The Informational City*. Oxford: Blackwell.

Castells, M. (1996) *The Rise of the Network Society*. Oxford: Blackwell.

Castells, M. (2001) *The Internet Galaxy. Reflections on the Internet, Business, and Society*. Oxford: Oxford University Press.

Chaiklin, S. and Lave, J. (1993) *Understanding Practice: Perspectives on Activity and Context*. Cambridge: Cambridge University Press.

Chandler, A.D. (1962) *Strategy and Structure*. Cambridge, MA: MIT Press.

Chandler, A.D. (1977) *The Visible Hand*. Cambridge, MA: Belknap.

Chandler, A.D. (1990) *Scale and Scope. The Dynamics of Industrial Capitalism*. Cambridge, MA: Harvard University Press.

Chandler, A.D., Hagstrom, P. and Solvell, O. (1998) *The Dynamic Firm. The Role of Technology, Strategy, Organization, and Regions*. Oxford: Oxford University Press.

Child, J. (1997) 'Strategic choice in the analysis of action, structure, organisations and environment: retrospect and prospect', *Organization Studies* 18 (1): 43–76.

Child, J. (2000) 'Theorizing about organization cross-nationally', *Advances in Comparative Management* 13: 27–75.

Child, J. and Loveridge, R. (1991) *Information Technology in European Services*. Oxford: Blackwell.

Child, J. and Smith, C. (1987) 'The context and process of organizational transformation: Cadbury Limited in sector', *Journal of Management Studies* 24: 563–593.

Clark, K.B. (1985) 'The interaction of design hierarchies and market concepts in technology evolution', *Research Policy* 14: 235–251.

Clark, K.B. and Fujimoto, T. (1991) *Product Development Performance. Strategy,*

Organization and Management in the World of Auto Industry. Cambridge, MA: Harvard Business School.

Clark P.A. (1972) *Organizational Design: Theory and Practice.* London: Tavistock.

Clark P.A. (1975) 'Key problems in organization design', *Administration and Society* 7: 213–256.

Clark, P.A. (1976) *Study of Time III.* Berlin: Springer-Verlag.

Clark, P.A. (1979) 'Cultural content as a determinant of organizational rationality: an empirical investigation of the tobacco industries in Britain and France', in C.J. Lammers and D.J. Hickson (eds), *Organizations: Alike and Unlike.* London: Macmillan.

Clark, P.A. (1985) 'A review of the theories of time and structures for organizational sociology', *Research in the Sociology of Organizations* 4: 35–79.

Clark, P.A. (1986) 'Le capitalisme et le règlement du temps de travail: une critique de la thèse E.P. Thompson', *Temps Libre* 15: 27–32.

Clark, P.A. (1987) *Anglo-American Innovation.* New York: De Gruyter.

Clark, P.A. (1990) 'Corporate chronologies and organisational analysis', in J. Hassard and D. Pymm (eds), *The Theory and Philosophy of Organisations.* London: Croom Helm.

Clark, P.A. (1997) 'American corporate timetabling, its past, present and future', *Time and Society* 6 (2/3): 261–285.

Clark, P.A. (2000) *Organizations in Action: Competition between Contexts.* London: Routledge.

Clark, P.A. and Mueller, F. (1996) 'Organisations and nations: from universalism to institutionalism?', *British Journal of Management* 7 (2): 125–140.

Clark, P.A. and Newell, S. (1993) 'Societal embedding of production and inventory control systems: American and Japanese influences on adaptive implementation in Britain', *International Journal of Human Factors in Manufacturing* 3: 69–81.

Clark, P.A. and Probert, S. (1989) 'The American tufted carpet revolution and two established British carpet firms – 1950/1985', paper to European Group on Organization Studies, Berlin.

Clark, P.A. and Starkey, K. (1988) *Organisation Transitions and Innovation Design.* London: Frances Pinter.

Clark, P.A. and Staunton, N. (1989) *Innovation in Technology and Organisation.* London: Routledge. (2nd Edition, 1993).

Clark, P.A. and Whipp, R. (1984) 'Industrial change in Britain: a comparative historical enquiry into innovation design', ESRC, Work Organization Research Centre, Aston University.

Clark, P.A., Carter, C. and Szmigin, I.T. (2000) The Spectrum of (Explacit) Knowledges in Firms and Nations. *Prometheus,* Vol 18–4. 454–460.

Clark, P.A., Dumas, A. and Hills, P. (1995) 'Design and technology: increasing the output through innovation in organisation', in D. Bennett and F. Steward (eds), *Technology Innovation and Global Challenges.* Birmingham: Aston University.

Clark, P.A., Newell, S., Swan, J., Bennett, D., Burcher, P. and Sharifi, S. (1992–93) 'The decision episode framework and computer-aided production management', (CAPM) *International Studies in Management and Organization* 22 (4): 69–80.

Clark, T. and Salaman, G. (1996) 'The management guru as organizational witch doctor', *Organization* 3: 85–108.

Clegg, S.R. (1989) *Frameworks of Power.* London: Sage.

Clegg, S.R. (1994) 'Weber and Foucault: social theory for the study of organizations', *Organization* 1: 149–178.

Clegg, S.R., Hardy, C. and Nord, J. (1996) *Handbook of Organisation Studies.* London: Sage.

Cohen, I.J. (1989) *Structuration Theory: Anthony Giddens and the Constitution of Social Life.* New York: St Martin's Press.

Cohen, M. and Bacdayan, P. (1994) 'Organizational routines are stored as procedural memory: evidence from a laboratory study', *Organization Science* 5 (4): 554–568.

Cohen, M.D., Burkhart, R., Dosi, G., Egidi, M., Marengo, L., Warglien, M. and Winter, S. (1996) 'Routines and other recurring action patterns of organizations: contemporary research issues', *Journal of Industrial and Corporate Change*, pp. 653–686.

Cohen, M.D., March, J.G. and Olson, J.P (1972) 'A garbage can model of organization choice', *Administrative Science Quarterly* 17: 1–25.

Cohen, W.M. and Levinthal, D.A. (1990) 'Absorptive capacity: a new perspective on organisational learning and innovation', *Administrative Science Quarterly* 55: 28–152.

Coleman, J.S. (1990) *Foundations of Social Behavior.* Cambridge, MA: Harvard University Press.

Cook, S.D.N. and Brown, J.S. (1999) 'Bridging epistemologies: the generative dance between organizational knowledge and organizational knowing', *Organization Science* 10 (4): 381–400.

Coombs, R., Hull, R. and Peltu, M. (1998) 'Knowledge management practices of innovation: an audit tool for improvement', Working Paper, CRIC-6. Manchester: University of Manchester.

Corke, D. (1985) *A Guide to Computer-Aided Production Management (CAPM).* London: Institute of Production Engineers.

Cornford, J. (1999) 'The virtual university is . . . the university made concrete?', CURDS Working Paper. Newcastle: University of Newcastle.

Cowan, R.S. (1987) 'The Consumption Junction: A Proposal for Research Strategies in the Sociology of Technology. In Bijker et al., pp. 261–280.

Crane, A. (2000) 'Corporate greening as amoralization', *Organization Studies* 21 (4): 673–696.

Crang, M. (1998) *Cultural Geography.* London: Routledge.

Cronon, W. (1983) *Changes in the Land. Indians, Colonists, and the Ecology of New England.* New York: Hill & Wang.

Cronon, W. (1991) *Nature's Metropolis: Chicago and the Great West.* New York: Norton.

Crozier, M. (1964) *The Bureaucratic Phenomenon.* Chicago, IL: University of Chicago Press.

Crozier, M. and Friedberg, E. (1980) *Actors and Systems.* Chicago, IL: University of Chicago Press.

Cusumano, M.A. (1985) *The Japanese Automobile Industry. Technology and Management at Nissan and Toyota.* Cambridge, MA: Harvard University Press.

Cyert, R.M. and March, J.G. (1963) *A Behavioral Theory of the Firm.* Englewood Cliffs, NJ: Prentice-Hall.

Czarniawska, B. (1998) *A Narrative Approach to Organisation Studies.* London: Sage.

Czarniawska, B. and Jorges, M. (1996) 'Travels of ideas'. In B. Czarniawska and G. Sevon (eds), *Translating Organizational Change*, pp. 13–47. Berlin: de Gruyter.

Daft, R. (1997) *Organization Theory and Design.* St Paul, MN: West.

David, P.A. (1975) *Technological Choice, Innovations and Economic Growth:*

Essays on American and British Experience in the 19th Century. Cambridge: Cambridge University Press.

Davies, H. and Ellis, P. (2000) 'Porter's competitive advantage of nations: time for the final judgement?', *Journal of Management Studies* 37 (8): 1189–1214.

Deleuze, G. (1992) 'Postscript on the societies of control', *October 59*, Winter: 3–7.

Demarest, M. (1997) 'Understanding knowledge management', *Journal of Long Range Planning* 30 (3): 374–384.

DeSanctis, G. and Poole, M.S. (1994) 'Capturing the complexity in advanced technology use: adaptive structuration theory', *Organization Science* 5: 121–147.

de Soto, H. (2000) *The Mystery of Capital. Why Capitalism Triumphs in the West and Fails Everywhere Else.* London: Bantam.

Di Bello, L., Kindred, J. and Zazanis, E. (1992) *Third Annual Report. Cognitive Studies of Work. Laboratory for Cognitive Studies of Activity.* New York: City University of New York.

Dicken, P. (1998) *Global Shift. Transforming the World Economy.* London: Paul Chapman.

DiMaggio, P.J. and Powell, W.W. (1983) 'The iron cage revisited: institutional isomorphism and collective rationality in organisational fields', *American Sociological Review* 48: 147–160.

DiMaggio, P.J. and Powell, W.W. (1991) 'Introduction', in W. Powell and P. DiMaggio (eds), *The New Institutionalism in Organisational Theory.* Chicago, IL: University of Chicago Press, pp. 3–45.

Djelic, M.L. (1998) *Exporting the American Model. The Postwar Transformation of European Business.* Oxford: Oxford University Press.

Dodd, N. (1995) 'Whither Mammon? Postmodern economics and passive enrichment', *Theory, Culture and Society* 12 (2): 1–24.

Doel, M. (1995) *Poststructuralist Geographies. The Diabolical Art of Spatial Science.* Edinburgh: Edinburgh University Press.

Dore, R.P. (1973) *British Factory: Japanese Factory.* London: Allen and Unwin.

Dosi, G. (1984) *Technical Change and Industrial Transformation.* New York: St Martin's Press.

Dower, J. (1999) *Embracing Defeat. Japan in the Aftermath of World War II.* London: Penguin.

Drazin, R. (1990) 'Professionals and innovation: structural–functional versus radical–structural perspectives', *Journal of Management Studies* 27 (3): 245–263.

Dubinskas, F.A. (1988) *Making Time. Ethnography of High-Technology Organizations.* Philadelphia, PA: Temple.

Dunning, J.H. (1993) *The Globalization of Business.* London: Routledge.

Elias, N. (1994) *The Civilising Process. The History of Manners and State Formation and Civilisation.* Oxford: Blackwell.

Elias, N. and Dunning, E. (1971) 'Folk football in medieval and early modern Britain', in E. Dunning (ed.), *Readings in the Sociology of Sport.* London: Frank Cass, pp. 313–327.

Emery, F.E. (1977) *Futures we are in.* Leiden: Martinus Nijhof.

Engestrom, Y. and Mietkinen, R. (????) (Eds) *Perspectives on Activity Theory.* Cambridge: Cambridge University Press.

Eveland, J.D. (1979) 'Issues in Using the Concept of "Adoption" in Innovations', *Journal of Technology Transfer* 4,1: 1–14.

Faust, M. (1999) 'The increasing contribution of management consultancies to management knowledge: the relevance of arenas for the communicative validation of knowledge', EGOS Working Paper. Tübingen: Tübingen University.

Ferguson, N. (ed.) (1997) *Virtual History: Alternatives and Counterfactuals.* London: Picador.

Firat, A.F. and Dholakia, N. (1998) *Consuming People: From Political Economy to Theatres of Consumption.* London: Routledge.

Fligstein, N. (1987) 'The intraorganizational power struggle: the rise of finance presidents in large corporations', *American Sociological Review* 52: 44–58.

Fligstein, N. (1990) *The Transformation of Corporate Control.* Cambridge, MA: Harvard University Press.

Fligstein, N. (1996) 'Markets as politics: a political-cultural approach to market institutions', *American Sociological Review* 61: 656–673.

Ford, D., Gadde, L.E., Lundgren, A., Snehota, I., Turnbull, P. and Wilson, D. (2000) *Managing Business Relationships.* Chichester: Wiley.

Foucault, M. (1972) *The Archaeology of Knowledge.* London: Tavistock.

Foucault, M. (1977) *Discipline and Punish: The Birth of the Prison.* Harmondsworth: Penguin.

Foucault, M. (1984) *The History of Sexuality: An Introduction.* Harmondsworth: Penguin.

Freeman, C. (1983) *Longwaves in the World Economy.* London: Butterworth.

Freeman, C. and Soete, L. (eds) (1985) *Technical Change and Full Employment.* Oxford: Blackwell.

Fujimoto, T. (1994) *Work Organization, Production Technology and Product Strategy of the Japanese and British Automobile Industries.* Commack, NY: Nova Science.

Fukuyama, F. (1992) *The End of History and the Last Man.* New York: Free Press.

Fukuyama, F. (1995) *Trust: The Social Virtues and the Creation of Prosperity.* New York: Free Press.

Galbraith, J.K. (1967) *The New Industrial Estate.* London: Hamish Hamilton.

Gandy, O.H. (1993) *The Panoptic Sort. A Political Economy of Personal Information.* Boulder, CO: Westview Press.

Garud, R. and Rappa, M.A. (1994) 'A socio-cognitive model of technology evolution: the case of cochlear implants', *Organization Science* 5: 344–362.

Gearing, E. (1958) 'Structural poses of the eighteenth-century Cherokee villages', *American Anthropologist* 60: 1148–1157.

Gereffi, G. (1994) 'Capitalism, development and global commodity chains', in L. Sklair (ed.), *Capital and Development.* London: Routledge, pp. 211–231.

Gereffi, G. and Korzeniewicz, M. (eds) (1994) *Commodity Chains and Global Capitalism.* Westport, CT: Greenwood Press, pp. 1–14.

Ghoshal, S. and Bartlett, C.A. (1994) 'Linking organizational context and managerial action: the dimensions of quality of management', *Strategic Management Journal* 15: 91–112.

Gibbons, M., Nowotny, H., Schwartzman, S., Scott, P. and Trow, M. (1994) *The New Production of Knowledge. The Dynamics of Science and Research in Contemporary Societies.* London: Sage.

Giddens, A. (1981) *A Contemporary Critique of Historical Materialism.* London: Macmillan.

Giddens, A. (1984) *The Constitution of Society.* Cambridge: Polity Press.

Giddens, A. (1990) *The Consequences of Modernity.* Cambridge: Polity Press.

Giddens, A. (1991) *Modernity and Self-identity: Self and Identity in the Late Modern Age*. Cambridge: Polity Press.

Giddens, A. (1994) *Beyond Left and Right*. Cambridge: Polity Press.

Giddens, A. (1998) *The Third Way*. Cambridge: Polity Press.

Gille, B. (1978) *Histoire des Techniques*. Paris: Piriade.

Glennie, P. and Thrift, N. (1996) 'Reworking E.P. Thompson's "Time, Work-Discipline and Industrial Capitalism"', *Time and Society* 5 (3): 275–299.

Grabher, G. (1993) *The Embedded Firm: On the Socioeconomics of Industrial Networks*. London: Routledge.

Granovetter, M.S. (1973) 'The strength of weak ties', *American Sociological Review* 78: 1360–1380.

Granovetter, M.S. (1985) 'Economic action and social structure: the problem of embeddedness', *American Journal of Sociology* 91: 481–510.

Granovetter, M. and Swedberg, R. (eds) (1992) *The Sociology of Economic Life*. Oxford: Westview Press.

Grant, R.M. (1990/1995/1998) *Contemporary Strategy Analysis. Concepts, Techniques and Applications*. Oxford: Blackwell.

Gregory, D. (1982) *Regional Transformation and Industrial Revolution. A Geography of the Yorkshire Woollen Industry*. London: Macmillan.

Gregory, D. (1994) *Geographical Imaginations*. Oxford: Blackwell.

Griffith, T.L. and Northcraft, G.B. (1994) 'Distinguishing between the forest and the trees: media, features and methodology in electronic communication research', *Organization Science* 5: 272–285.

Grint, K. and Woolgar, S. (1998) *The Machine at Work. Technology, Work and Organization*. Cambridge: Polity Press.

Grove, G. and Fisk, J. (1983) *Interactive Services Marketing*. Boston, MA: Houghton Mifflin.

Gruneau, R.G. (1999) *Class, Sports and Social Development*. Champagne, IL: Leeds.

Guillen, M. (1994) *Models of Management: Work, Authority, and Organization in a Comparative Perspective*. Chicago, IL: University of Chicago Press.

Gurvitch, G. (1964) *The Spectrum of Social Time*. Dordrecht: Reidel.

Gutek, B.A., Bikson, T.K. and Mankin, D. (1984) 'Individual and organizational consequences of computer-based office information technology', in S. Oskamp (ed.), *Applied Social Psychology Annual: Applications in Organizational Settings*. Beverly Hills, CA: Sage, pp. 231–254.

Hage, J. and Hollingsworth, J.R. (2000) 'A strategy for the analysis of idea innovation networks and institutions', *Organization Studies* 21 (5): 971–1004.

Hagerstrand, T. (1978) 'Survival and arena', in Carlstein et al. (eds), *Timing Space and Spacing Time*. London: Edward Arnold, pp. 128–145.

Hakansson, H. (ed.) (1988) *Industrial Technology Development. A Network Approach*. London: Croom Helm.

Hall, J.R. (1992) 'Where history and sociology meet: forms of discourse and sociohistorical inquiry', *Sociological Theory* 10 (2): 164–193.

Hall, R.H. (1972) *Organisations: Structure and Process*. Englewood Cliffs, NJ: Prentice-Hall.

Hammer, M. and Champy, J. (1993) *Re-engineering the Corporation. A Manifesto for Business Revolution*. London: Harper Collins.

Hannan, M. and Freeman, J. (1977) 'The population ecology model of organisations', *American Journal of Sociology* 82: 929–964.

Hannan, M. and Freeman, J. (1989) *Organisational Ecology*. Cambridge, MA: Harvard University Press.

Hardy, S. (1982) *How Boston Played: Sport, Recreation and Community*. Boston, MA: Northeastern University Press.

Harré, R. (1979) *Social Being: A Theory for Social Psychology*. Oxford: Blackwell.

Harré, R. (1985) *Varieties of Realism*. Oxford: Blackwell.

Harvey, D. (1989) *The Condition of Postmodernity*. Oxford: Blackwell.

Hassard, J. (1996) 'Images of time in work and organization', in S. Clegg, J. Nord and C. Hardy (eds), *Handbook of Organisation Studies*. London: Sage, pp. 381–309.

Hasselbladh, H. and Kallinikos, J. (2000) 'The project of rationalization: a critique and re-appraisal of Neo-Institutionalism in Organization Studies', *Organization Studies* 21 (4): 697–720.

Hatch, M.J. (1997) *Organisation Theory*. Oxford: Oxford University Press.

Haunschild, P.R. and Miner, A.S. (1997) 'Modes of interorganizational imitation: the effects of outcome salience and uncertainty', *Administrative Science Quarterly* 42: 472–500.

Haveman, H.A. (1993) 'Follow the leader: mimetic isomorphism and entry into new markets', *Administrative Science Quarterly* 38: 593–627.

Hawthorne, G. (1991) *Plausible Worlds: Possibility and Understanding in History and the Social Sciences*. Cambridge: Cambridge University Press.

Henderson, R.M and Clark, K.B. (1990) 'Architectural innovation: the reconfiguration of existing product technologies and the failure of established firms', *Administrative Science Quarterly* 35: 9–30.

Henderson, R. and Cockburn, I. (1994) 'Measuring core competence? Evidence from the pharmaceutical industry', *Strategic Management Journal* 15: 63–84.

Hickson, D.J. and Pugh, D.S. (1995) *Management Worldwide*. Harmondsworth: Penguin.

Hickson, D.J., Hinings, C.R., McMillan, C.J. and Schwitter, J.P. (1974) 'The culture-free context of organisation structure: a tri-national comparison', *Sociology* 8: 59–80.

Hippel, E. von (1988) *The Sources of Innovation*. Oxford: Oxford University Press.

Hirsch, P.M. and Lounsbury, M.D. (1996) 'Rediscovering volition: the institutional economics of Douglas C. North', *Academy of Management Review* 21: 872–884.

Hirsch, P.M. (1997) 'Sociology without social structure: Neo-institutional theory meets brave new world', *American Journal of Sociology* 102: 1702–1723.

Hislop, D., Newell, S., Scarbrough, H. and Swan, J. (1997) 'Innovation and networks: linking diffusion and implementation', *International Journal of Innovation Management* 1 (4): 427–448.

Hodgson, G.M. (1988) *Economics and Institutions: A Manifesto for a Modern Institutional Economics*. Cambridge: Polity Press.

Hodgson, G.M. (1999) *Economics and Utopia: Why the Learning Economy is not the End of History*. London: Routledge.

Hofstadter, R. (1955) *Social Darwinism in American Thought*. Boston, MA: Beacon Press.

Hofstede, G. (1980) *Culture's Consequences: International Differences in Work-related Values*. Beverly Hills, CA: Sage.

Hollingsworth, J.R. and Boyer, R. (eds) (1997) *Contemporary Capitalism: The Embeddedness of Institutions*. New York: Cambridge University Press.

Holt, D.B. (1997) 'Post-structuralist lifestyle analysis: conceptualizing the social patterning of consumption in postmodernity', *Journal of Consumer Research* 23–1: 216–350.

Hoskin, K.W. and Macve, R. (1988) 'The genesis of accountability: the West Point connections. Accounting', *Organizations and Society* 13: 37–73.

Hounshell, D.A. (1984) *From the American System to Mass Production, 1800–1932*. Baltimore, MD: Johns Hopkins University Press.

Hudson, R. (1999) 'The learning economy, the learning firm and the learning region: a sympathetic critique of the limits to learning', *European Urban and Regional Studies* 6 (1) 59–72.

Huff, A. (2000) 'Changes in organizational knowledge production', *Academy of Management Review* 25 (2): 288–293.

Hughes, A. (2000) 'Retailers, knowledges and changing commodity networks: the case of the cut flower trade', *Geoforum* 31: 175–190.

Hughes, T.P. (1983) *Networks of Power: Electrification in Western Society*. Baltimore, MD: Johns Hopkins University Press.

Hughes, T.P. (1987) 'The Evolution of Technological Systems'. In Bijker et al., pp. 51–82.

Hughes, T.P. (1990) *American Genesis*. Baltimore, MD: Johns Hopkins University Press.

Hutchins, E. (1993) 'Learning to navigate', in S. Chaiklin and J. Lave (eds), *Understanding Practice. Perspectives on Activity and Context*. Cambridge: Cambridge University Press.

Jameson, F. (1991) *Postmodernism, or, The Cultural Logic of Late Capitalism*. London: Verso.

Jeremy, D.J. (1981) *Transatlantic Industrial Revolution*. Oxford: Blackwell.

Johansson, H.J., McHugh, P., Pendlebury, A.J. and Wheeler, W.A. (1993) *Business Process Re-engineering*. New York: Wiley.

Joint Information Services Committee (1996) *JISC Information Strategies Initiative*. Bristol: JISC.

Jones, C. (2000) 'Strategic networks in American film, 1895–1920', Working Paper. Boston, MA: Boston College.

Kagan World Media (2000) London: Paul Kagan Associates.

Khedia, B.L. and Bhagat, R.S. (1988) 'Cultural constraints on transfer of technology across nations: implications for research in international and comparative management', *Academy of Management Review* 13 (4): 559–571.

Kieser, A. (1997) 'Rhetoric and myth in management fashion', *Organization* 4 (1): 49–74.

Kennedy, P. (1987) *The Rise and Fall of Great Powers: Economic Change and Military Conflict 1500–2000*. New York: Random House.

Kenney, M. and Florida, R. (1993) *Beyond Mass Production. The Japanese System and its Transfer to the United States*. Oxford: Oxford University Press.

Kipping, M. and Bjarnar, O. (1998) *The Americanization of European Business*. London: Routledge.

Kline, S. and Rosenberg, N. (1986) 'An overview of innovation', in R. Landau and N. Rosenberg (eds), *The Positive Sum Strategy*. Washington, DC: National Academy Press, pp. 275–305.

Kogut, B. (1993) 'Country capabilities and the permeability of borders', *Strategic Management Journal* 12: 33–47.

Kogut, B. and Zander, U. (1992) 'Knowledge of the firm, combinative capabilities and the replication of technology', *Organization Science* 3 (3): 383–397.

Kogut, B. and Zander, U. (1996) 'What firms do?: Coordination, identity and learning', *Organization Science* 7 (5): 502–519.

Kuhn, T.S. (1970) *Structure of Scientific Revolutions*. Chicago: Chicago University Press.

Lakatos, I. (1978) 'Falsification and methodology of scientific research programs', in I. Lakatos and A. Musgrave (eds), *Criticism and the Growth of Knowledge*. Cambridge: Cambridge University Press.

Lammers, C.J. and Hickson, D.J. (eds) (1979) *Organisations Alike and Unlike: International and Inter-institutional Studies in the Sociology of Organisations*. London: Macmillan.

Landes, D. (1998) *The Wealth and Poverty of Nations*. London: Little Brown.

Larson, M.S. (1977) *The Rise of Professionalism. A Sociological Analysis*. Berkeley, CA: University of California Press.

Lash, S. (1990) *Sociology of Postmodernism*. New York: Routledge.

Lash, S. (1999) *Another Modernity. A Different Rationality*. Oxford: Blackwell.

Lash, S. and Urry, J. (1987) *End of Organized Capitalism*. Cambridge: Polity Press.

Lash, S. and Urry, J. (1994) *Economy of Signs and Spaces*. London: Sage.

Latour, B. (1987) *Science in Action: How to Follow Scientists and Engineers Through Society*. Milton Keynes: Open University Press.

Latour, B. (1988) *The Pasteurization of France*. Cambridge: Cambridge University Press.

Latour, B. (1991) 'Technology is society made durable', in J. Law (ed.), *A Sociology of Monsters: Essays on Power, Technology and Domination*. London: Routledge.

Latour, B. (1993) *We Have Never Been Modern*. Cambridge, MA: Harvard University Press.

Lave, J. and Wenger, E. (1991) *Situated Learning. Legitimate Peripheral Participation*. Cambridge: Cambridge University Press.

Law, J. (1986) 'On the methods of long-distance control: vessels, navigation and the Portuguese route to India', in J. Law (ed.), *Power, Action and Belief: A New Sociology of Knowledge? Sociological Review Monograph 32*. London: Routledge and Kegan Paul, pp. 234–83.

Law, J. (1994) *Organising Modernity*. Oxford: Blackwell.

Lears, J. (1994) *Fables of Abundance. A Cultural History of Advertising in America*. New York: Basic Books.

Lefebvre, H. (1974/1991) *The Production of Space*. Oxford: Blackwell.

Legge, K. (1995) *Human Resource Management: Rhetorics and Realities*. London: Macmillan.

Levinthal, D.A. and March, J.G. (1993) 'The myopia of learning', *Strategic Management Journal* 14: 95–112.

Leyshon, A. and Thrift, N. (1997) *Money/Space. Geographies of Monetary Transformation*. London: Routledge.

Lillrank, P. (1995) 'The transfer of management innovations from Japan', *Organization Studies* 16 (6): 971–989.

Lipset, S.M. (1996) *American Exceptionalism. A Double-edged Sword*. New York: Norton.

Locke, R.R. (1996) *The Collapse of the American Management Mystique*. Oxford: Oxford University Press.

Lockwood, D. (1964) 'Social integration and system integration', in G.K. Zollschan and W. Hirsch (eds), *Explorations in Social Change*. London: Routledge & Kegan Paul.

Lowe, M. and Wrigley, N. (1999) *Reading Retail*. London: Arnold.

Lundvall, B.A. (1992) *National Systems of Innovation. Towards a Theory of Innovation and Interactive Learning*. London: Pinter.

Lupton, T. (1964) *Management and the Social Sciences*. London: Penguin.

Lury, C. (1996) *Consumer Culture*. Cambridge: Polity Press.

Lyon, D. (2001) *Surveillance Society*. Milton Keynes: Open University Press.

Lyotard, J.F. (1984) *The Postmodern Condition: A Report on Knowledge*. Minneapolis, MN: University of Minnesota Press.

MacKenzie, D. (1990) *Inventing Accuracy: An Historical Sociology of Ballistic Missile Guidance*. Cambridge, MA: MIT Press.

Maguire, J. (1999) *Global Sport: Identities, Societies, Civilizations*. Oxford: Polity Press.

Mahoney, T.T. and Pandian, R. (1998) 'Resource-based view within the conversation of strategic management', *Strategic Management Journal* 13: 363–380.

Mandel, E. (1980) *Longwaves of Capitalist Development: The Marxist Interpretation*. Cambridge: Cambridge University Press.

March, J.G. (1991) 'Exploitation and exploration in organizational learning', *Organization Science* 2 (1): 71–87.

Marshall, A. (1961) *Principles of Economics*. London: Macmillan.

Marx, K. (1970) *Capital*. London: Lawrence & WIshart.

Marx, K. (1976) *Grundrisse*. Harmondsworth: Penguin.

Mathias, K. and Bjarnar, O. (eds) (1998) *The Americanisation of European Business. The Marshall Plan and the Transfer of US Management Models*. London: Routledge.

Matusik, S. and Hill, C.W.L. (1998) 'The utilization of contingent work, knowledge creation and competitive advantage', *Academy of Management Review* 23: 680–697.

Mauss, M. (1904) 'Essai sur les variations saisonniers des sociétés Eskimaux', *L'Année Sociologique*, Volume IX.

McMillan, C.J. (1985) *The Japanese Industrial System*. Berlin: de Gruyter.

McNay, I. (1995) 'From the collegial academy to corporate enterprise: the changing cultures of universities', in T. Schuller (ed.), *The Changing University?* Buckingham: Open University Press/SRHE, pp. 105–115.

Meyer, J.W. and Rowan, B. (1977) 'Institutionalized organizations: formal structure as myth and ceremony', *American Sociological Review* 83: 340–363.

Miles, M.B. and Huberman, A.M. (1994) *Qualitative Data Analysis*. London: Sage.

Miller, D. (1997a) *Capitalism. An Ethnographic Approach*. Oxford: Berg.

Miller, D. (ed) (1997b) *Material Cultures*. London: University College Press.

Miller, D. (1998) *A Theory of Shopping*. Cambridge: Polity Press.

Miller, D. and Slater, D.R. (2000) *The Internet. An Ethnographic Approach*. Oxford: Berg.

Mintzberg, H. (1978) *The Structuring of Organizations. A Synthesis of the Research*. Englewood Cliffs, NJ: Prentice Hall.

Mizruchi, M.S. and Fein, L.C. (1999) 'The social construction of organizational knowledge: a study of the uses of coercive, mimetic and normative isomorphism', *Administrative Science Quarterly* 44: 653–683.

Mohr, L.B. (1982) *Explaining Organisational Behavior: The Limits and Possibilities of Theory and Research*. San Francisco, CA: Jossey-Bass.

Mokyr, J. (1990) *The Lever of Riches. Technological Creativity and Economic Progress*. Oxford: Oxford University Press.

Monden, Y. (1981) 'What makes the Toyota production system really tick?', *Industrial Engineering* 13 (1): 36–46.

Montgomery, C.A. (1995) *Resource-based and Evolutionary Theories of the Firm: Towards a Synthesis.* Cambridge, MA: Harvard Business School.

Nahapiet, J. and Ghoshal, S. (1998) 'Social capital, intellectual capital and the organizational advantage', *Academy of Management Review* 23 (2): 242–266.

Nelson, R.R. (1991) 'Why do firms differ, and how does it matter?', *Strategic Management Journal* 12: 61–74.

Nelson, R.R. (ed.) (1993) *National Innovation Systems.* Oxford: Oxford University Press.

Nelson, R.R. and Winter, S.G. (1977) 'In search of useful theory of innovation', *Research Policy* 6 (1): 36–77.

Nelson, R.R. and Winter, S.G. (1982) *An Evolutionary Theory of Economic Change.* Cambridge, MA: Harvard University Press.

Newell, S. and Clark, P.A. (1990) 'The importance of extra-organization networks in the diffusion and appropriation of new technologies. The role of professional associations in the United States and Britain', *Knowledge: Creation, Diffusion, Utilization* 12 (2): 199–212.

Newell, S., Swan, J. and Clark, P.A. (1993) 'The importance of user design in the adoption of new information technologies. The example of production and inventory control systems', *Journal of Production and Operations Management* 13 (2): 1–22.

Newell, S., Swan, J. and Robertson, M. (1997) 'Exploring the diffusion of BPR across manufacturing firms in Europe', Proceedings of the 5th European Conference on Information Systems. Cork: Cork Publishing Company.

Newell, S., Scarbrough, H. and Swan, J. (2001) 'From global knowledge management to internal electronic fences: contradictory outcomes of intranet development', *British Journal of Management* 12 (2): 97–112.

Newman, K.L. (2000) 'Organizational transformation during institutional upheaval', *Academy of Management Review* 25 (3): 602–619.

Noble, D. (1978) *American by Design.* New York: Knopf.

Noble, D. (1984) *Forces of Production. A Social History of Industrial Automation.* New York: Knopf.

Noble, D.F. (1998) 'Digital diploma mills: the automation of higher education', *Science as Culture* 7 (3): 355–368.

Nonaka, I. and Takeuchi, H. (1995) *The Knowledge Creating Company. How Japanese Companies Create the Dynamics of Innovation.* Oxford: Oxford University Press.

North, D. (1990) *Institutions, Institutional Change, and Economic Performance.* New York: Cambridge University Press.

Nowotny, H., Scott, P. and Gibbons, M. (2001) *Re-thinking Science. Knowledge and the Public in an Age of Uncertainty.* Cambridge: Polity Press.

Nye, D.E. (1997) *Narratives and Spaces. Technology and the Construction of American Culture.* Exeter: Exeter University Press.

Oberschall, A. and Leifer, E.M. (1986) 'Efficiency and social institutions: uses and misuses of economic reasoning in sociology', *Annual Review of Sociology* 12: 233–53.

Oliver, C. (1991) 'Strategic responses to institutional processes', *Academy of Management Review* 16: 145–179.

Oliver, C. (1992) 'The antecedents of deinstitutionalisation', *Organization Studies* 13 (4): 653–588.

References

Oliver, C. (1997) 'Sustainable competitive advantage: Combining institutional and resource based views', *Strategic Management Journal* 18: 607–713.

O'Malley, M. (1990) *Keeping Watch: A History of American Time*. New York: Viking Penguin.

O'Neill, P. and Whatmore, S. (2000) 'The business of place: networks of property, partnerships and produce', *Geoforum* 31: 121–136.

Orlicky, J. (1977) *Material Requirements Planning*. New York: McGraw Hill.

Orlikowski, W.J. (1992) 'The duality of technology: rethinking the concept of technology in organizations', *Organization Science* 3: 398–427.

Orlikowski, W.J. (2000) 'Using technology and constituting structures: A practice lens for studying technology in organizations', *Organization Science* 4: 404–428.

Orlikowski, W.J. and Gash, D.C. (1994) 'Technological frames: making sense of information technology in organizations', *ACM Transactions of Information Systems* 12: 174–207.

Orlikowski, W.J. and Robey, D. (1991) 'Information technology and the structuring of accounting, management and information technology', *Organisations, Information System Research* 2 (2): 143–169.

Orlikowski, W.J. and Yates, J. (2000) 'It's About Time. Temporal Structuring in Organizations', *Working Paper*. Cambridge, MA: Sloan School of Management, MIT.

Orlikowski, W.J., Yates, J., Okamura, K. and Fujimoto, M. (1995) 'Shaping electronic communication: the metastructuring of technology in the context of use', *Organization Science* 6: 423–444.

Parker, M. and Jarry, D. (1995) 'The McUniversity: organisation, management and academic subjectivity', *Organization* 2 (2): 319–338.

Parkes, D.N. and Thrift, N.J. (1980) *Times, Spaces and Places. A Chronographic Perspective*. Chichester: Wiley.

Parsons, T. (1951) *The Social System*. New York: Free Press.

Pascale, R. (1990) *Managing on the Edge. How Successful Companies Use Conflict to Stay Ahead*. London: Penguin.

Pavitt, K. (1980) *Technical Change and Britain's Economic Performance*. London: Pinter.

Penrose, E.T. (1959) *The Theory of the Growth of the Firm*. Oxford: Blackwell.

Perrow, C. (1967) 'A framework for the comparative analysis of organizations', *American Sociological Review* 32: 194–208.

Pettigrew, A.M. (1973) *The Politics of Organizational Decision Making*. London: Tavistock.

Pettigrew, A.M. (1985) *The Awakening Giant. Continuity and Change in ICI*. Oxford: Blackwell.

Philo, G. and Miller, D. (2001) *Market Killing. What the Free Market Does and What Social Scientists Can Do About It*. London: Longman.

Pine, B.J. (1993) *Mass Customization. The New Frontier in Business Competition*. Cambridge, MA: Harvard Business School.

Pollock, N. and Cornford, J. (2000) 'Theory and practice of the Virtual University', *Ariadne* (24), 21 June.

Porter, M.E. (1990) *The Competitive Advantage of Nations*. New York: Free Press.

Porter, M.E. (1998) 'Clusters and Competition: New Agendas for Companies, Governments, and Institutions', in M.E. Porter (ed.), *On Competition*. Cambridge, MA: Harvard Business Review Book.

Porter, M.E., Takeuchi, H. and Sakakibara, M. (2000) *Can Japan Compete?* London: Macmillan.

Powell, W.W. (1990) 'Neither market nor hierarchy: network forms of organization', in B. Staw and L.L. Cummings (eds), *Research in Organizational Behavior*. Greenwich, CT: JAI Press, pp. 295–336.

Powell, W.W. and DiMaggio, P.J. (1991) *The New Institutionalism in Organization Analysis*. Chicago, IL: University of Chicago Press.

Powell, W.W., Koput, K.W. and Smith-Doerr, L. (1996) 'Interorganizational collaboration and the locus of innovation: networks of learning in biotechnology', *Administrative Science* Quarterly 41: 116–145.

Power, M. (1994) *The Audit Explosion*. London: DEMOS.

Pugh, D.S. and Hickson, D.J. (1976) *Aston Program*. London: Saxon House.

Pulos, A.J. (1983) *American Design Ethic. A History of Industrial Design to 1940*. Cambridge, MA: MIT Press.

Putnam, R.D. (1993) *Making Democracy Work. Civic Traditions in Modern Italy*. Princeton, NJ: Princeton University Press.

Putnam, R.D. (2000) *Bowling Alone. The Collapse and Revival of American Community*. New York: Simon & Schuster.

Ritzer, G. (1998) *The McDonaldization Thesis: Explorations and Extensions*. London: Sage.

Rogers, E.M. (1962) *Diffusion of Innovations*. New York: Free Press (2nd edition, 1983; 3rd edition, 1995).

Rogers, E.M. and Rogers, R.K. (1976) *Communication in Organizations*. New York: Free Press.

Rosenberg, N.A. (1969) *The American System of Manufacturing*. Edinburgh: Edinburgh University Press.

Resenberg, N.A. and Steinmuller, W.E. (1988) 'Why are Americans such poor imitators?', *American Economic Association Papers and Proceedings*, pp. 229–234.

Roth, J.A. (1963) *Timetables*. Indianapolis: Bobbs-Merrill.

Roudometof, V. and Robertson, R. (2001) *Nationalism, Globalization & Orthodoxy*. New Jersey: Greenwood.

Rowlinson, M. (1977) *Organizations and Institutions*. London: Macmillan.

Rugman, A.M. (1995) *Research in Global Strategic Management, Vol. 5: Beyond the Diamond*. Greenwich, CT: JAI Press.

Ryan, Y. (2000) 'The business of borderless education: US case studies and the HE response', Paper presented to the CVCP Conference – The Business of Borderless Education – London, 28 March.

Sahay, S. (1998) 'Information systems in organizations: time–space perspective', *Organization Studies* 18 (2): 229–260.

Said, E.W. (1994) *Culture and Imperialism*. London: Vintage.

Sayer, A. (1992) *Method in Social Science*. London: Routledge.

Sayer, A. (2000) *Realism and Social Science*. London: Sage.

Scarbrough, H. (1995) 'Blackboxes, hostages and prisoners', *Organization Studies* 16 (6): 991–1020.

Scarbrough, H. (1998) 'Path(ological) dependency? Core competencies from an organizational perspective', *British Journal of Management* 9: 219–232.

Scarbrough, H. and Swan, J. (2001) 'Explaining the diffusion of knowledge management: the role of fashion', *British Journal of Management* 12: 3–12.

Schama, S. (1991) *Dead Certainties. (Unwarranted Speculations)*. London: Granta.

Schonberger, R.J. (1981) *Japanese Manufacturing Techniques. Nine Hidden Lessons in Simplicity*. New York: Free Press.

Schonberger, R.J. (1996) *The Cultural Crisis of the Corporations*. Oxford: Oxford University Press.

Schudson, M. (1984) *Advertising, the Uneasy Profession*. New York: Basic Books.

Schultz, M., Hatch, M.J. and Larson, M.H. (2000) *The Expressive Organization. Linking Identity, Reputation and the Corporate Brand*. Oxford: Oxford University Press.

Schumpeter, J.A. (1942/1975) *Capitalism, Socialism, and Democracy*. New York: Harper & Row.

Scott, W.R. (1995) *Institutions and Organizations*. London: Sage.

Scribner, S. (1997) *Mind and Social Practice: Selected Writings*. Cambridge: Cambridge University Press.

Searle, J.R. (1995) *The Construction of Social Reality*. London: Penguin.

Senge, P.M. (1990) *The Fifth Discipline*. New York: Doubleday.

Sennett, R. (1998) *The Corrosion of Character. The Personal Consequences of Work in the New Capitalism*. New York: Norton.

Sklair, L. (2000) *The Transnational Capitalist Class*. Oxford: Blackwell.

Slater, D.R. (1997) *Consumer Culture and Modernity*. Cambridge: Polity Press.

Slater, D.R. and Tonkiss, F. (2001) *Market Society. Markets and Modern Social Theory*. Cambridge: Polity Press.

Soja, E.W. (1996) *Third Space. Journeys to Los Angeles and Other Real-and-Imagined Places*. Oxford: Blackwell.

Sorge, A. (1991) 'Strategic fit and the societal effect: interpreting cross-national comparisons of technology, organization and human resources', *Organization Studies* 12: 161–90.

Sorge, A. (1995) 'Cross-national differences in personnel and organisation', in A.W. Harzing and J.V. Ruysseveldt (eds), *International Human Resource Management*. London: Sage, pp. 99–123.

Sorkin, M. (1992) *Variations on a Theme Park: The New American City and the End of Public Space*. Los Angeles: Hill and Wang.

Star, L.A. and Ruhdler, K. (1996) 'Steps toward an ecology of infrastructure: design and access for large information spaces', *Information Systems Research* 7 (1): 111–134.

Storper, M. and Salais, R. (1997) *World of Production: The Action Frameworks of the Economy*. Cambridge, MA: Harvard University Press.

Stymne, B. (ed.) (1996) *International Transfer of Organizational Innovations*. Stockholm: IMIT Publications.

Suchman, L. (1987) *Plans and Situated Actions. The Problem of Human–Machine Communication*. Cambridge: Cambridge University Press.

Swan, J. and Clark, P.A. (1992) 'Organization decision-making in the appropriation of technological innovation: political and cognitive dimensions', *European Work and Organizational Psychologist* 2 (2): 103–127.

Swan, J. and Newell, S. (1999) 'Central agencies in the diffusion and design of technology: a comparison of the UK and Sweden', *Organization Studies* 20 (6): 905–931.

Swan, J., Newell, S. and Robertson, M. (1999) 'National differences in the diffusion and design of technological innovation: the role of . . .', *British Journal of Management* 10 (3): S45–S59.

Szmigin, I.T.D. (1997) 'Cognitive style and the use of payment methods: an interpretive study', Doctoral thesis, Birmingham: University of Birmingham.

Szmigin, I.T. (2003 forthcoming) *Understanding the Consumer*. London: Sage.

Thevenot, L. (1984) 'The investment in forms', *Social Science Information* 23 (1): 1–45.

Thevenot, L. and Boltanski, L. (1988) *Economies de Grandeur*. Paris: Presses Universitaires de France.

Thompson, E.P. (1967) 'Time, work-discipline and industrial capitalism', *Past and Present* 35: 56–97.

Thompson, J.D. (1967) *Organisations in Action*. New York: Wiley.

Thrift, N. (1983) 'On the determination of social action in space and time', *Environment and Planning D, Society and Space* 1: 23–57.

Thrift, N. (1996) *Spatial Formations*. London: Sage.

Thrift, N. (1997) 'The rise of soft capitalism', *Cultural Values* I: 27–57.

Tidd, J. and Fujimoto, T. (1994) 'Work organization, production technology and product strategy of the British and Japanese automobile industries', *Current Politics and Economics of Japan* 4 (4): 241–281.

Torres, R. (2000) 'Image commodification and the evolution of taste: the case of food in the global economy', Working Paper, Department of Rural Sociology. Ithaca, NY: Cornell University Press.

Townley, B. (1994) *Reframing Human Resource Management*. London: Sage.

Trist, E.L., Higgin, G.W., Murray, H. and Pollock, A.B. (1963) *Organizational Choice: the Loss, Re-discovery and Transformation of a Work Tradition*. London: Tavistock.

Tsoukas, H. (1994) 'What is management? An outline of a metatheory', *British Journal of Management* 5: 289–301.

Turnbull, D. (2000) *Masons, Tricksters and Cartographers*. Amsterdam: Harwood.

Tyre, M.J. and Orlikowski, W.J. (1994) 'Windows of opportunity: temporal patterns of technological adaptation in organizations', *Organization Science* 5: 98–118.

Urry, J. (1991) 'Time and space in Giddens' social theory', in C. Bryant and D. Jary (eds), *Giddens' Theory of Structuration*. London: Routledge, pp. 160–175.

Urry, J. (1995) *Consuming Places*. London: Routledge.

Van de Ven, A.H. (1986) 'Central problems in the management of innovation', *Management Science* 32 (5).

Van de Ven, A.H. and Poole, M.S. (1995) 'Explaining development and change in organizations', *Academy of Management Review* 20.3, 510–540.

Van de Ven, A.H., Angle, H.L. and Poole, M.S. (1989) *Research on the Management of Innovation: The Minnesota Studies*. New York: Harper & Row.

Von Bertalanffy, L. (1968) *General System Theory: Foundations, Developments, Applications*. New York: Braziller.

Wallerstein, I. *The Modern World-System II: Mercantilism and the Consolidation of the World-Economy, 1600–1750*. New York: Academic Press.

Weber, M. (1922/1978) *Economy and Society: An Outline of Interpretive Sociology*. Berkeley, CA: University of California Press.

Weber, M. (1923/1982) *General Economic History*. Translated F.H. Knight. New Brunswick, NJ: Transaction Books.

Weber, M. (1947) *The Theory of Social and Economic Organization*. London: Routledge.

Weick, K.E. (1969) *Social Psychology of Organising*. New Haven, CT: Addison-Wesley (second edition, 1979).

Weick, K.E. (1995) *Sensemaking in Organisations*. London: Sage.

Weick, K.E. (2001) *Making Sense of the Organization*. Oxford: Blackwell.

References

Wenger, E. (1998) *Communitites of Practice: Learning, Meaning and Identity.* Cambridge: Cambridge University Press.

Whipp, R. and Clark, P.A. (1986) *Innovations and the Automobile Industry: Product, Process and Work Organization.* London: Pinter.

Whitley, R.D. (1994) 'Dominant forms of economic organization in market economies', *Organization Studies* 15 (2): 153–182.

Whitley, R.D. (1998) 'Internationalism and the varieties of capitalism: the limited effects of cross-national coordination of economic activities on the nature of business systems', *Review of International Political Economy* 5 (3): 445–481.

Whittington, R. (1992) *What is Strategy and Does it Make a Difference?* London: Sage.

Whittington, R. and Mayer, M. (2000) *The European Corporation. Strategy, Structure and Social Science.* Oxford: Oxford University Press.

Whittington, R., Bouchikhi, H. and Kilduff, M. (1995) *Action, Structure and Organisations.* Coventry: Warwick Business School Research Bureau.

Wight, O. (1984) *MRPII: Unlocking America's Productivity Potential.* Willeston, VT: Wight Press.

Wight, O. (1985) *Manufacturing Resource Planning: MRPII.* Willeston, VT: Oliver Wight.

Williams, R. (1985) *Culture and Society 1780–1950.* London: Harmondsworth.

Williamson, O.E. (1985) *The Economic Institutions of Capitalism.* New York: Macmillan.

Wilson, D.C. (1992) *A Strategy of Change.* London: Routledge.

Winter, S.G. (1990) 'Survival, selection, and inheritance in evolutionary theories of organization', in J.V. Singh (ed.), *Organizational Evolution: New Directions.* London: Sage.

Wolfe, R.A. (1994) 'Organizational innovation: review, critique and suggested research directions', *Journal of Management Studies* 31 (3): 405–431.

Woodward, J. (ed.) (1970) *Industrial Organization: Behaviour and Control.* London: Oxford University Press.

Woolgar, S. (1981) 'Interests and explanation in the social study of science', *Social Studies of Science* 11: 365–94.

Wrigley, N. and Lowe, M. (eds) (1996) *Retailing, Consumption and Capital: Towards the New Retail Geography.* Harlow: Longman.

Yates, J. (1989) *Control Through Communication: The Rise of American Management.* Baltimore, MD: Johns Hopkins University Press.

Yates, J. and Orlikowski, W.J. (1992) 'Genres of organizational communication: a structural approach to studying communication and the media', *Academy of Management Review* 17 (2): 299–327.

Zaltman, G. and Duncan, R. (1977) *Strategies for Planned Change.* New York: Wiley.

Zeitlin, J. (1995) 'Americanization and its limits: Theory and practice in the reconstruction of Britain's engineering industries 1945–1955', *Business and Economic History* 24 (1): 277–286.

Zeldin, T. (1994) *An Intimate History of Humanity.* London: Minerva Press.

Zuboff, S. (1988) *The Age of the Smart Machine.* New York: Basic Books.

Index

NOTE: Page numbers in *italic type* refer to tables or figures.

.